T0296196

The Terry Lectures

ORDER AND LIFE

ORDER AND LIFE

BY

JOSEPH NEEDHAM

FELLOW OF GONVILLE AND CAIUS COLLEGE
SIR WM. DUNN READER IN BIOCHEMISTRY
UNIVERSITY OF CAMBRIDGE

All things began in Order, so shall they end, and so shall they begin again; according to the Ordainer of Order, and the mysticall Mathematicks of the City of Heaven.

SIR THOMAS BROWNE, *The Garden of Cyrus* (1658), ch. 5.

CAMBRIDGE · AT THE UNIVERSITY PRESS

New Haven · Yale University Press

1936

CAMBRIDGE
UNIVERSITY PRESS

University Printing House, Cambridge CB2 8BS, United Kingdom

Cambridge University Press is part of the University of Cambridge.

It furthers the University's mission by disseminating knowledge in the pursuit of education, learning and research at the highest international levels of excellence.

www.cambridge.org
Information on this title: www.cambridge.org/9781107504837

First published 1936
First paperback edition 2015

A catalogue record for this publication is available from the British Library

ISBN 978-1-107-50483-7 Paperback

TO THE

THEORETICAL BIOLOGY CLUB

J.H.W.

C.H.W. D.W.

M.B. J.D.B.

B.P.W. D.M.N.

E.W.

CONTENTS

ILLUSTRATIONS

INTRODUCTION

IN introducing a series of lectures such as these, it is proper for an author to express his sense of the great honour done him by Yale University in inviting him to give them. It is a pleasure to me to do this as warmly as possible.

But there may be other sentiments of a general character which require expression. In the present case I have to confess to certain misgivings as to the appropriateness of the subject I have chosen, namely, the consideration of what some would call the cleavage between the "inorganic" and the biological sciences, but what I should prefer to call the cleavage between morphology and biophysics or biochemistry. In the realm of scientific order there are cracks or "faults," as a geologist would say, and it is certainly worth while to make a serious attempt to see just how deep-lying and far-reaching such faults are. But it was only after a good deal of thought that I convinced myself that this subject was appropriate for a Terry Lecturer.

The deed of gift of the Terry Lectures runs as follows: "The object of this Foundation is not the promotion of scientific investigation and discovery, but rather the assimilation and interpretation of that which has been or shall be hereafter discovered, and its application to human welfare, especially by the building of the truths of science and philosophy into the structure of a broadened and purified religion."

Accordingly I feel it necessary to say at the outset that in what follows the first object of the Terry Foundation will be under discussion, and that whatever conclusions can be drawn regarding the second will be implicit rather than explicit. We shall be engaged with the "assimilation and interpretation" of what has hitherto been discovered, particularly in the attempt to inquire whether the universe of science is really one, or whether there remain within it any wholly unbridgeable chasms.

To any who may feel disappointment that distinctively theological issues are not taken up, I would recommend the words of Sir Thomas Browne: "Thus there are two books from whence I collect my Divinity; besides that written one of God, another of his servant Nature, that universal and publick Manuscript, that lies expans'd unto the Eyes of all: those that never saw him in the one, have discover'd him in the other. This was the Scripture and Theology of the Heathens; the natural motion of the Sun made them more admire him than its supernatural Station did the children of Israel; the ordinary Effects of Nature wrought more admiration in them than in the other all his Miracles. Surely the Heathens knew better how to joyn and read these mystical Letters than we Christians, who cast a more careless Eye on these common Hieroglyphicks, and disdain to suck Divinity from the flowers of Nature."[1]

Some, of course, will not set so high a value upon "natural theology." They will regard the proposi-

1. *Rel. Med.* (1643), Pt. 1, §16.

tions of theism as philosophically barren, even logically meaningless, and will prefer a purely naturalistic account of creation followed by a dialectical production of infinite manifoldness out of primitive homogeneity. But to them also my problem presents itself—the "mathematicks" of empirical order, and the question of its unity.

As for the others, it is notable that two main ways exist whereby the truth of a disputed belief may be demonstrated. There is firstly the argument from *inexplicability*, which consists in saying that the thing in question is so odd or so unique or so different from anything else we know that the mind could never have arrived at it unless it were true. Sir Thomas Browne, we find, represents this opinion too when he says: "I can answer all the Objections of Satan and my rebellious reason with that odd resolution I learned of Tertullian—*Certum est quia impossibile est.*"[2] It is so extraordinary that it must be true. This line of argument, represented with great force in Huxley's brilliant essay on Job,[3] leads directly to the realm of mystical theology. It is associated with many of the greatest names in the history of religious thought, with Chrysostom, Irenæus, Nicholas of Cusa, William James, Rudolf Otto. It deserves the greatest respect.

But there is a second way—the argument from *intelligibility*, which consists in showing how well the

2. *Loc. cit.*, §9.

3. Aldous Huxley, in *Do What You Will* (London, Chatto & Windus).

thing in question fits in with the rest of our knowledge or our opinions about the universe. Thus rational theology has had a history running always parallel with that of mystical theology. The Kantian revolution, which demolished the old "proofs of the existence of God," ended by bringing him back in order to explain the moral law. Belief in God was called in to explain the experience of values rather than the facts of nature, and this situation is still with us. Of course, the extreme predilection for rational theology found in some historical periods led to results at least as unfortunate as uncontrolled mysticism, as in the eighteenth century, when the Deists gave excellent reasons for believing in God but supplied no God to be believed in.

We see in historical method the operation of these two arguments.[4] Scholars discussing the credibility of an ancient historian could use either. The fact reported might be held to be so extraordinary and unlike what we should have expected that the historian is credible because he could never have invented anything so unlikely. On the other hand, it might be exactly what we would predict from the rest of our knowledge of the time, and then the historian is credible because there was no need for him to invent anything so likely. In so far, then, as the discussion which I shall give about the oneness of inorganic and organic science has any theological implications, it will be because the intelligibility of the universe is

4. They are well described in O. Quick, *The Ground of Faith and the Chaos of Thought* (London, Nisbet, 1931).

increased by such oneness. The essential business of constructive science is to increase the intelligibility of the universe. This process the rational theologian will welcome. Although the mystical theologian will not perhaps actually deprecate it, he may not be very interested in it. It is not for me to say which of them is right, although I will not deny that I have my suspicions.

Lastly, I cannot refrain from adding that a famous series of Gifford Lectures was entirely devoted to introducing a certain biological phantom into the (usually) polite society of theology. I have felt it my duty to devote a part of the present lectures to withdrawing and cancelling such credentials as the Entelechy has had. There is no future for it in theoretical biology, and theology is the home of a quite sufficient number of lost causes already.

I
THE NATURE OF BIOLOGICAL ORDER

THOSE who are accustomed to ponder the ultimate problems of biology are aware that, though the need for a comprehensive biological science is great, the difficulties in obtaining it are equally considerable. Such old antitheses as that of form and function need not, indeed, detain us, for, as Woodger's analysis once and for all made clear,[1] form is simply a short time-slice of a single spatio-temporal entity. We have enlarged our concept of structure so as to include and recognise that in the living organism there is no spatial structure with an activity as something over against it, but that the concrete organism is a spatio-temporal structure and that this spatio-temporal structure is the activity itself. But the fusion of the two great realms of morphology and biochemistry, or, if we take the more all-inclusive term, biophysics, still remains. What is the relation between those large particles which we call elephants, trees, or men, and those extremely small ones which we call molecules or electrons? Because at the present day the biochemist has often little enough to offer toward the solution of the problem of the origin and maintenance of organic form, the morphologist is apt to suppose that no connections exist, and to

1. Woodger, J. H., *Biological Principles* (London, Kegan Paul, 1929), chap. vii.

acquiesce in an acceptance of the ancient Aristotelian distinction between *materia* and *forma*. This, however, is a counsel of despair.

At the outset it is necessary to emphasize that the traditional concepts of biological mechanism and vitalism will not come in for much consideration. Here the turning-point was reached with the fruitful suggestion, made by Woodger in 1929,[2] that for the future the term "vitalism" should be restricted to theories which postulate some entity in the living organism *in addition to* the chemical elements, C, H, N, O, P, etc., plus organising relations. This formulation was only another way of asserting that organisation exists in the living organism, and that this organisation is not something fundamentally mystical and unamenable to scientific attack, but rather the basic problem confronting the biologist.[3]

The perennial service of old-fashioned vitalism in

2. Woodger, J. H., *loc. cit.*

3. It may be noted that while some sorts of organising relations may be peculiar to living organisms, not all of them are. We may certainly speak of the organising relations in a soap bubble, or a simple crystal. In these cases we may know a good deal about them, but in other cases, such as, for example, the cholesteric type of paracrystal (*see* p. 160), they are almost as obscure as the organising relations in the living cell itself. It is often forgotten that mere complicatedness is not what characterises living systems. Many physical problems may be unsolved because of their complicatedness; thus it is impossible to find the electric capacity of a solid except in the simplest cases, e.g., ellipsoids. And again the nature of biological organising relations is only relevant to the biophysicist, just as an aerodynamician must work with the relevant concepts, but need know nothing about valency or wave-mechanics. This is an application of the distinction between parameters and functions (cf. the discussion below on the "autonomy" of biology, p. 19).

all its forms was that it continually drew attention to the real complexity of the phenomena, and opposed the tendency, so common among the mechanists, of putting forward over-simplified hypotheses. The new point of view ensures that we shall not forget the extreme complicatedness of the living system, while at the same time forbidding us to take refuge in pseudo-explanations. The older vitalism could hardly be acquitted of leanings toward romantic animism, it *hoped* that rigid causal analysis would fail, whereas mechanism hoped it would win. If science wins, as Marvin[4] said, the world will prove to be one in which man is thrown wholly on his own resources, skill, and self-control, on his courage and strength and perhaps on his ability to be happy by adjusting himself to pitiless fact. If science fails there is room for the childlike hope that unseen powers may come to the relief of human weakness. If science wins, the world is the necessary consequence of logically related facts, and man's enterprise, in Huxley's figure of speech, the playing of a game of chess against an opponent who himself never errs and never overlooks our errors. If science fails, the world is a fairyland and man's enterprise no longer a task for skill and knowledge, but conditioned by the goodness of his will and the possibility of luck. The motivation of biological mechanism was thus progressive, vigorous, and youthful, a seeking for independence and mastery. Vitalism had more affinity with the religious attitude

4. Marvin, W. T., "Mechanism and Vitalism as a Philosophical Issue," *Philos. Rev.*, *27* (1918), 616.

of creaturely dependence upon a higher power, and in its emphatic affirmation of the complexity of the phenomena of life manifested something of that numinous respect for the otherness in things, which properly belongs to religious experience. It is to be noted that the new conception of biological organisation combines the insistence of vitalism on the real complexity of life with the heuristic virtues of the mechanistic practical attack.

As an illustration of these characteristics of vitalism we may study the thought of that *doyen* of physiologists, J. S. Haldane.[5] Biologists recently read his Donnellan lectures[6] with that extraordinary mixture of admiration for the author and exasperation at his views which they have now long been accustomed to experience. As Berkeley inquired whether there is any meaning in the physical universe apart from its being perceived, so Haldane wishes to inquire whether any meaning can be attached to the concept of an organism apart from its environment. It is obvious, he says, that in the biological sciences we are always making use of words such as "function," "organ," "species," which are quite peculiar to biology and have no analogy in the physico-chemical sciences. These words cover a residuum (the major part of biology) for which no physico-chemical de-

5. Or rather "vitalistic organicism," or "neo-vitalism," for Haldane has always carefully repudiated older presentations as well as the views of Driesch.

6. Haldane, J. S., *The Philosophical Basis of Biology; Donnellan Lectures in the University of Dublin, 1930* (London, Hodder & Stoughton, 1931).

scription has ever been found. Indeed, "the more we discover as to physiological activity and inheritance, the more difficult does it become to imagine any physical or chemical description or explanation which could in any way cover the facts of persistent coordination."[7] We cannot separate organic from environmental structure, for no sharp line of demarcation can be drawn between organism and environment. So we come to Haldane's axiom: "The active maintenance of normal and specific structure is what we call life, and the perception of it is the perception of life. The existence of life as such is thus the axiom upon which scientific biology depends."

It is almost profitless to criticise this point of view, for scarcely any modification has taken place in Haldane's opinions since the days of his paper in Andrew Seth's collaborative *Essays in Philosophical Criticism* of 1883.[8] His brilliant physiological researches have been carried on side by side with an attitude to biological phenomena which can only be likened to that of Fracastorius of Verona—"The heart's motion God alone can comprehend." Time after time Haldane appeals to the arguments from presumed impossibility and inconceivability. With respect to organism and environment, we easily see that his demand for the unification of the organism and its surroundings is a methodologically impossible

7. *Loc. cit.*, p. 12.

8. Haldane, R. B. and J. S., "The Relation of Philosophy to Science," in *Essays in Philosophical Criticism*, ed. A. Seth (later Pringle-Pattison) and R. B. Haldane (London, 1883).

aim, for if no line can be drawn between organism and immediate surroundings, no better line can be drawn between immediate surroundings and far-off surroundings. Biology is thus brought under the rule, not only of Berkeley, but also of Plotinus, and we are invited to contemplate the universe in its axiomatic wholeness, analysis of living things being laid aside.

But the solid foundation for Haldane's position, the real burden of his emphasis reiterated during so many years, is of course that the problem of organisation is the central problem of biology, and that the riddle of form is the fundamental riddle. "When we discover, for instance," he says, "the existence of an intraprotoplasmic enzyme or other substance on which life depends, we are at the same time faced with the question how this particular substance is present at the right time and place, and reacts to the right amount to fulfill its normal functions."[9] Or again: "Physical and chemical mechanisms within the living body are actively maintained at the right places and in the right functional states."[10] This is the first step in his reasoning, and it is one which cannot be gainsaid, for it refers to facts of universal observation. But his second step is to confer upon this biological organisation an axiomatic or arbitrary character, to make it something which we must accept as given.

Now that there is arbitrariness in the universe

9. *Donellan Lectures*, p. 79. 10. *Loc. cit.*, p. 16.

cannot be disputed. Why the universe has the nature it does have, and not some other nature, is not a question to which any scientific answer can be given, although it may have import for teleology and theology. The alogical core of the world, as Tennant well says, is not a residuum of haze which science, when ideally perfect, shall have dissipated, it is on the contrary a determinateness, occult, inexplicable, and rationally incomprehensible.[11] But the worst mistake that bad scientific methodology can make is to introduce this element at a higher level than is absolutely necessary. The whole essence of scientific systematisation is the attempt to render intelligible the primary chaos of arbitrary entities. This we see very clearly in a young science such as that of comparative biochemistry. The main outlines and much of the detail of comparative morphology have long been familiar, but in the special biochemistry of the phyla all is yet to do, and we can as yet give no rational answer to such questions as why the tunicates should contain curious vanadium compounds in their blood,[12] or why the pycnogonids should possess a special pigment unknown anywhere else in the animal kingdom.[13] Only by long and persistent comparative researches, pursued always with reference to œcological

11. Tennant, F. R., *Philosophical Theology* (Cambridge, University Press, 1928–30), Vol. I, chap. xiii.

12. Henze, M., "Uber das Vanadiumchromogen des Ascidienblutes," *Zeitschr. f. physiol. Chem., 213* (1932), 125.

13. Dawson, A. B., "The Coloured Corpuscles of the Blood of the Purple Sea-spider *Anoplodactylus lentus*," *Biol. Bull., 66* (1934), 62.

and other conditions of life, will a rational schema emerge. To accept any of our present knowledge as axiomatic or directly related to the alogical core of the universe would be precisely not the way to advance our knowledge of the subject.

The minimisation of the alogical core of the universe, then, is the proper pursuit of science. The older parts of biology, such as comparative morphology, are still faced with the apparently insoluble problem of why there should be a given number of phyla and no more. It is unlikely that the time has yet come for any satisfactory answer to this type of question, but there seems no reason for supposing that such a time will never come. The attempts of Galton and Clark[14] to show that there are no "missing links" between the phyla, but only between the groups within them, and their suggestion that the phyla represent so many positions of morphological stability into which primordial living matter could fall, though not perhaps very convincing, are nevertheless a typical example of the kind of theory that the scientific mind does and must produce in its struggle against the arbitrary aspect of the universe. At an earlier stage, the consideration of the possible as well as the actual is an important methodological tool, for no reason can be advanced in favour of a given alternative unless the other alternatives are known, and unless some reason is found, the process remains in the obscurity of its arbitrary limbo. "We

14. Clark, A. H., *The New Evolution, Zoogenesis* (London, Baillère, Tindall & Co, 1930).

need scarcely add," says Eddington, "that the contemplation in natural science of a wider domain than the actual leads to a far better understanding of the actual."[15] Fisher,[16] commenting on this, urges that biologists should work out the detailed consequences which would be experienced by organisms having three or more sexes. And he goes on: "The ordinary mathematical procedure in dealing with any actual problem is, after abstracting what are believed to be its essential elements, to consider it as one of a series of possibilities infinitely wider than the actual, the essentials of which may be apprehended by generalised reasoning and subsumed in general formulae which may be applied at will to the particular case considered." The bearing of this on modern attempts to rationalise comparative physiology is obvious. In considering possible types of tetrapod, a relation is found between corpuscular haemoglobin and the organisation of mammals,[17] or in considering the conquest of terrestrial habitat, we find a connection between the terrestrial oviparous animals and uricotelic metabolism.[18] Here we are investigating the limiting factors which, as it were, select the actual from the row of candidate possibles.

Thus to ask how living organisms come to have the

15. Eddington, A. S., *The Nature of the Physical World* (Cambridge, University Press, 1928), p. 266.

16. Fisher, R. A., *Genetical Theory of Natural Selection* (Oxford, 1930), p. viii.

17. Barcroft, J., "The Significance of Submammalian Forms of Haemoglobin," *Physiol. Rev. 4* (1924), 329, *5* (1925), 596.

18. Needham, J., "Protein Metabolism and Organic Evolution," *Sci. Progr., 23* (1929), 633.

properties they do have is to ask what other properties they could have had. And here we border upon that reciprocal relation of fitness between organism and environment which has been treated of (to borrow a phrase which Matthew Slade used of William Harvey) in "that golden book" of Lawrence Henderson's, *The Fitness of the Environment.*[19] Here, as will be familiar to all, the evolutionary process was shown to be not a matter of chance, but inevitable, granting the general principles of biological organisation, and the properties of the chemical elements—a conclusion at least as acceptable to dialectical materialism as to orthodox theology. Vitalism was thus dissolved in universal teleology, but again we see that the alogical core, far from being abolished, is brought into clearer illumination, and we still have the question why this unified evolutionary process should have the characteristics it actually did. This happens however naturalistic your account of cosmogony. However, it is clear that although the minimisation of the alogical core may be an asymptotic process, it is none the less an essential part of scientific thought. With this we return to biology proper and to Haldane's conception of biological organisation as arbitrary.

If we take biological organisation as axiomatic, that is, as an essential part of the impenetrable alogical core, we at once remove it from the realm of experiment. We are adopting a course as fruitless as it

19. Henderson, L. J., *The Fitness of the Environment* (New York, Macmillan, 1913).

would be to take the chemical elements as axiomatic in physics and to make no attempt to go behind the periodic table into the realm of the electronic structure of atoms. If this had been the *non plus ultra* of physical chemistry we should never have heard anything of electronic orbits in atoms of different elements, and the radioactive transformation of one element into another would have remained an inscrutable mystery. The sublime expression "I am that I am" is well suited to the manifestation of a deity, but when applied to the immediate problems confronting scientific workers, its use becomes nothing more than the frank confession of intellectual bankruptcy.

Yet such confessions on the part of eminent biologists are in our time disturbingly frequent. In what follows I hope to show that the obscure relations between morphological form and chemical change are open to experimental attack at many points. But there is a tendency to regard the problem as impossibly difficult and to postpone its consideration until the Greek kalends. "The biologist," says J. Gray, "must accept the living state as he finds it." "It seems logical to accept the existence of matter in two states, the animate and the inanimate, as an initial assumption."[20] Such a point of view can hardly be

20. Gray, J., "The Mechanical View of Life," address in *The Advancement of Science: 1933* (London, British Association, 1933), p. 86, (abbreviated in *Nature, 132* [1933], 661). Similar statements are quite frequent. Thus E. S. Russell says: "The fact of the unity, in a spatio-temporal sense of the organism, constitutes not so much a problem as a postulate." (*The Interpretation of Development and Heredity,* Oxford, Clarendon Press, 1930,

conscious of its own sterilising power. "As I understand it," he goes on to say, "the age-long discussion between the mechanist and vitalist schools of thought turns on how far we believe, on the basis of observation, that the facts of biology can be sorted out into an harmonious and satisfying series without invoking conceptions which are found to be unnecessary in dealing with the facts of observation within the physical world."[21] On the contrary, as we have seen, the inclusion of the special type of organisation found in living systems within the sphere of science has nothing whatever to do with vitalism, which posits some entity in addition to organising relations. The "irreducibility" of biological categories can receive quite another interpretation, for the laws of the nematic or smectic state in liquid crystals (about which more will be said hereafter) are similarly irreducible to those holding good for common isotropic liquids. It is for us to investigate the nature of this

p. 6.) Also J. S. Haldane: "Life as simply life is the reality which must be assumed in biological interpretation" (*Materialism,* London, Hodder & Stoughton, 1932, p. 66). I myself at one time held opinions which, though very different from the vitalist ones entertained by these authors, led to the same conclusions. I regarded the nature of biological organisation as a purely philosophical question, and excluded it from scientific biology. The difference was that, unlike these authors, I did not maintain that the causal-analytical method of science could tell us little or nothing about the nature of life. I had not seen the full significance of the analogous science of crystallography. I am glad to have an opportunity of cancelling what I then said ("Organicism in Biology," *Journ. Philos. Stud., 3* [1928], 29, subsequently reprinted in *The Sceptical Biologist,* London, Chatto & Windus, 1928; New York, Norton, 1930).

21. Gray, J., *loc. cit.,* p. 82.

biological organisation, not to abandon it to the metaphysicians because the rules of physics do not seem to apply to it.

It is sometimes valuable to distinguish between "obstructionist" or "dogmatic" organicism, and "legitimate" organicism.[22] For the former point of view, organisation is quite inscrutable since, it is urged, any part instantly loses its relational properties on removal from the whole, and no means are available to us for making wholes transparent and inspecting them while still intact. Fortunately for biology, these propositions are not true. The question has been admirably discussed by Woodger,[23] who distinguishes three main possibilities in the relation of organic part to whole (a) independence, (b) functional dependence, (c) existential dependence. A part of the first sort would pursue its normal activities independently of whether it was in connection with its normal whole or not. A part of the second sort would not do so, and a part of the third sort would not even be recognisable when removed from its whole. Dogmatic organicists, ignoring these important distinctions, assume that all parts are parts of the third sort. Yet this is certainly not the case. Even existential dependence is a difficulty which can be overcome if means exist for making wholes "transparent." It is of much historical interest that a controversy quite similar to this was going on in the seventeenth century, only

22. As in the theoretical introduction to my *Chemical Embryology* (1931).

23. Woodger, J. H., *Proc. Aristot. Soc.*, *32* (1932), 117.

then with reference to inorganic mixtures as analysed by the chemists as well as what we should now call organic wholes.[24] The chemists defended themselves rationally enough.

There is a good deal more to be said about the "irreducibility" of biological categories, or the "autonomy" of biology, as it is sometimes called. Here the supreme guide is always Wilhelm Roux.[25] Although the biological philosophy of the founder of

24. Cf. the following passage in Nicholas Lemery's *A Course of Chymistry* (London, 1698), p. 6:

"Some modern Philosophers would perswade us, that it is altogether uncertain, whether the substances which are separated from Bodies, and are called Chymical Principles, do effectually exist and are naturally residing in the Body before: these do tell us that the Fire by rarifying the matter in time of distillation is capable of bestowing upon it such an alteration as is quite different from what it had before, and so of forming the Salt, Oil, and other things which are drawn from it. This objection does at first seem to have much weight and reason in it, because it is certain that the Fire does give a very considerable impression to the preparations, and that very often it does put such a new face upon things, that they are very hardly to be known when compar'd with what they were before. But it is easie to shew, that though Fire does so diversifie and alter substances, yet it does not make these Principles; for we see them and smell them in many bodies, before ever we bring them to undergo the Fire. For example, it cannot be denied, but that there was existent Oil in Olives, in Almonds, in Nuts, and in many other Fruits and Seeds, because it is drawn only by beating and pressing them. Turpentine, which is a thicken'd Oil, and many other fat, or unctuous liquors, are drawn by meer incision into the trunk or root of trees; and what else, I pray, is the fat of Animals but an Oil, or Sulphur coagulated? Nor can it be denied, but that there is a Salt actually in mixt Bodies, since by bruising a Plant, and making expression to draw out its juice, and then leaving the juice to settle in some cool place for a few days, a Salt will be found fixt about the vessel in form of little Crystals."

25. *See* his *Die Entwicklungsmechanik d. Organismen, eine anatomische Wissenschaft d. Zukunft* (Vienna, 1890).

Entwicklungsmechanik was thoroughly mechanistic, he nevertheless realised the difficulty of expounding the processes of development immediately in terms of physico-chemical concepts. His primary aim was to show that by the simple application of the causal concept much could be done, and he was content to leave open for the time being the question of the physico-chemical interpretation of the regularities which would thus be observed. "The too simple mechanistic conception on the one hand," he wrote in 1895, "and the metaphysical conception on the other, represent the Scylla and Charybdis, between which to sail is indeed difficult, and so far by few satisfactorily accomplished."[26]

Now the course he set, which took him through, involved the analysis of development, not into simple physico-chemical processes directly, but into more complex organic processes.[27] He showed that it was

26. *Archiv. f. Entwicklungsmechanik, 1* (1895), 23.

27. For a vigorous defence of Roux's methodology, cf. Hartmann, M., "Die methodologischen Grundlagen d. Biologie," *Erkenntnis, 3* (1932), 235, and *Nova Acta Leopoldina,* N.F. *1* (1933), 294, "Die biologischen Kausalgesetzlichkeiten in erster Linie spezifische Gesetze der Komplizierung sind." From the current number of the *Arch. f. Entwicklungsmech.* I choose at random an account of causal work on the complex components. Perri explants the eye-rudiment of amphibia, at the neurula stage and earlier, into a neutral culture medium. Under these conditions he finds it retains outside the body its properties of self-differentiation, regulation, fusion with a portion of another eye-cup, lens-induction in indifferent ectoderm, and lens-formation from the edge of the iris. It is clear that, unless the properties of such systems are well understood at the biological level, no ultimate success can attend the investigation of the finer (physico-chemical) levels. (Perri, T., *Arch. f. Entwicklungsmech., 131* [1934], 113.)

possible, and indeed necessary, to deal first with large packets of factors in the biological organism, before proceeding to the finer analysis of the smaller packets. The aim of *Entwicklungsmechanik* is thus the reduction of the phenomena to the smallest number of causal processes, *Wirkungsweisen*. The large packets of factors, or as he called them, the complex components, are not as yet explicable in physico-chemical terms; the smaller packets, or simple components, such as we have of them, are so explicable. So long as the complex components are found to be constant in their action, and always under the same conditions, to produce the same effect, causal biology would be on the right lines of analysis. These biological generalisations would thus be as valid as those of physics and chemistry, though possessing a more complex content. So long as the expression "biological autonomy" means no more than this, so long is it perfectly justified. But unfortunately it is often extended to cover the purely dogmatic view that the complex components never will be translatable into physico-chemical terms, whether of to-day or to-morrow. In this latter sense it is, of course, to be rejected.

The whole vast domain of experimental morphology which has arisen since Roux's time bears witness to the clarity and strength of his intuition. But there is sometimes now to be found a desire to replace the methods of experimental morphology by those of physiology in the hope of obtaining thereby a short cut into the arcana of biological organisation. This is to be wrecked on the Scylla of which he spoke. It is an impatience with the large amount of thinking that has still to be done on the purely biological level. Yet to suppose that this state of things can be avoided is a complete illusion. Experimental embryology, as can be seen from the admirable recent book of Huxley and de Beer,[28] has up to the present time mainly been concerned with the influence of part on part during embryonic development. In this way, by observing the behaviour of parts in abnormal situations, the whole array of organiser phenomena was discovered, and a great deal more knowledge based on transplantations brought into being. Empirical discoveries on the purely biological level thus serve as stimuli to the physiologist to investigate processes which his methods alone would never have revealed in the first place. For example, the concept of morphogenetic fields, to be referred to in more detail later, arises in the pure realm of complex components, and must there be thoroughly tested, clarified and defined, before any chemical analysis of their properties can begin. To characterise experimental embryology, therefore, as "one of the most backward branches of biological science"[29] (Wells) is to reveal a complete failure to understand the process of biological discovery.

28. *Elements of Experimental Embryology* (Cambridge, University Press, 1934).

29. Wells, G. P., "Morphogenesis in the Animal Embryo," *Nature, 133* (1934), 890.

This failure probably originates from a confusion of thought about the part which mathematics should play in science. "Physiological analysis," says Wells, "depends very greatly for its ideas and methods on physics and chemistry, and in these sciences the emphasis has been on quantities rather than on shapes." Is not this a false antithesis? Both shape and quantity must surely be regarded as ultimately definable in terms of numbers. A topological system or a figure in solid geometry is surely as numerical in its way as a finite number of quantitative weight-units. We are not indeed as yet equipped with the mathematical armamentarium which morphology requires, but this should not be allowed to obscure for us the fact that the central problem of biology is the form-problem. After all, it is essential to realise that, although the quantitative in the restricted sense of chemistry has a great part to play in biology, nevertheless simple arithmetic does not exhaust the realm of logical order, nor is it the only form which scientific exactness may take.[30] There are other systems of structure besides arithmetic, and the complex components may be very faithfully and logically dealt with on their own level. An outstanding example of this is the accuracy of prediction obtainable in modern genetics. And, as we have said, the complex components must

30. Cf. Nagel, E., "Measurement," *Erkenntnis, 2* (1931), 313. "Mathematics cannot be regarded as exclusively the science of quantity. Its essence is the study of types of order, of which the quantitative one is a single instance." Cf. Kohnstamm, P., *Journ. Phil. Studies, 5* (1930), 159.

be ordered in this way before they can be linked with physico-chemical knowledge.

The question of the mathematisation of morphology is one which is very urgently present in the minds of many biologists, both theoretical and practical. Thus Bertalanffy[31] discusses the famous saying of Kant that in any branch of natural knowledge the amount of true science present is directly proportional to the amount of mathematics. If the word mathematics is here taken in its strictest sense, the remark is nothing but a statement of one's conviction of the genuine *Rationalisierbarkeit* of Nature. Bertalanffy realises that there are strange realms in mathematics, unknown to most biologists, out of which concepts essential for the understanding of living systems may come, and can envisage the utilisation for biology of order-systems not even involving number and quantity. We see, he says, how in physics new groups of facts require the creation of new sorts of mathematics, as in the quantum theory, matrix theory, etc. The majority of biologists, on the other hand, suppose that the application of mathematics to biology means no more than the use of the calculus for curve-fitting, for example, in growth, or the working out of temperature coefficients (apparently the most variable constants known). But Bertalanffy's revolutionary dictum, which should be engraved on the hearts of all his readers, runs "Only

31. Bertalanffy, L. v., *Theoretische Biologie* (Berlin, Bornträger, 1932).

with the closest co-operation of the theoretical physicist, the mathematician, and the mathematical logician, will the problem of the mathematisation of biology be solved."

Other theoretical biologists, such as A. Meyer,[32] are also rightly very concerned to urge this point of view. Modern science, he says, must be mathematical or nothing. A causal science can, indeed, be nothing if not mathematical, since the ideal axiom at the basis of all causality can only be stated in terms of the mathematical concept of function. Physical equations necessarily involve functions. But an important distinction must be drawn between mathematical and mechanical. The modern ideal of a mathematical science was first brought into being in mechanics through the *Principia* of Newton. It was not surprising that mechanics should thus be taken as the perfect example of scientific systematisation, both present and to come. Other sciences were supposed to be going to develop, not merely *like* mechanics, but into special chapters *of* mechanics. Up to the end of the last century it was the universal assumption that it would some day be possible to refer all natural motions back to the basic equations of classical mechanics. But we know now that, far from sociology being referable to mechanics by way of biology and physics, physics itself has failed to conform to this artificial standard. Neither electrodynamics,

32. Meyer, A., *Ideen und Ideale der biologischen Erkenntnis* (Leipzig, Barth, 1934), p. 24.

atomic physics nor quantum theory can be derived from the principles of classical mechanics. But because biology cannot now ever be in the strict sense mechanistic, we have no right to assume that it cannot be causal, or that it cannot be mathematical.

Now Haldane, too, lays considerable emphasis on the fact that the assumed basic conceptions of Newtonian physics have been shown to be not in reality basic, and that instead we are presented with facts (for example, the intense co-ordinated specific activity within the atom, which does not become dissipated in its environment, and upon which its mass and other properties depend), which bear some resemblance to those with which biologists are familiar. This is certainly very important, but everything here depends on the precise degree of this resemblance. The comparison between the atom and the organism is in essence an analogy, and analogies are notoriously liable to snap under the weight placed upon them by uncautious thinkers. No one is yet in a position to say what the influence of the quantum theory and wave mechanics may be upon biology, but Haldane, curiously blind to these possibilities, can only infer from the breakdown of Newtonian physics the breakdown of physical and chemical biology. To more unbiassed observers, the vanishing crudeness of chemistry makes the future of biochemistry all the more hopeful.

The celebrated Haldanian dictum that "if a meeting-place between biology and physics be some day found, and one of the two sciences be swallowed up, that one will not be biology"[33] is thus supposed by many apocalyptists to be on the verge of fulfilment.[34] Bohr himself[35] has recently put forward some very rash observations upon the analogy mentioned above. "The existence of life," he says, "must be considered as an elementary fact that cannot be explained, but must be taken as a starting-point in biology, in a similar way as the quantum of action, which appears as an irrational element from the point of view of classical mechanical physics, taken together with the existence of the elementary particles, forms the foundation of atomic physics."[36] There is a sense in which this statement is platitudinous, but all its other senses are regrettably likely to discourage research into the nature of biological organisation. We have here a

33. Haldane, J. S., *The New Physiology* (London, Griffin, 1919), p. 19.

34. Thus Meyer (*loc. cit.*, pp. 73, 84, 147, 172, etc.), in his interesting discussion of the concept of wholeness, maintains that the fundamental conceptions of physics ought to be deducible from the fundamental conceptions of biology; the latter not being reducible to the former. Thus Entropy would be, as it were, a special case of biological disorganisation; the uncertainty principle would follow from the psycho-physical relation; and the principle of relativity would be derivable from the relation between organism and environment. The plausibility of this theme must be judged by recourse to Meyer's own text.

An analogous tendency may be seen in the contention of P. Kohnstamm (*Journ. Phil. Studies, 5* [1930], 159) that some parts of thermodynamics can best be regarded as the non-quantitative morphology of inorganic matter. Kohnstamm objects, as we have seen, to the view that exact scientific method is synonymous with measurement, for topology and analysis situs are non-metrical.

35. Bohr, N., "Light and Life," *Nature, 131* (1933), 421, 457. A somewhat more simple-minded presentation of the same notion occurs in E. S. London, *Archiv. f. mik. Anat., 97* (1923), 48.

36. *Loc. cit.*, p. 458.

mere assumption, supported by no single argument, the assumption that the form of order exemplified by the two realms is fundamentally identical.

From this first analogy Bohr proceeds to a second, in which the principle of uncertainty in physics is correlated with what may be called the thanatological principle in biology. It is now a matter of general knowledge that physicists have come to regard the exact determination of both the position and the velocity of a particle as impossible. They found that the more accurately they attempted to specify its position, the less accurately could the velocity or momentum be determined, and vice versa. This uncertainty is connected with the relation between the size of the electron and the wave-length of the light by which it might be observed. With long wave-lengths no exact definition can be obtained. When the wave-length is decreased enough to give definition, the radiation knocks the electron out of its position. It is characteristic of all the phenomena of light, says Bohr, in the description of which the wave picture plays an essential rôle, that any attempt to trace the paths of the individual light quanta would disturb the very phenomena under investigation; just as an interference pattern would completely disappear if, in order to make sure that the light energy travelled only along one of the two paths between the source and the screen, we introduced a non-transparent body into one of the paths. The spatial continuity of light-propagation, on the one hand, he goes on, and the atomicity of the light effects on the other hand, must therefore be considered as complementary aspects of one reality, in the sense that each expresses an important feature of the phenomena of light, which although irreconcilable from a mechanical point of view can never be in direct contradiction, since a closer analysis of one or the other feature in mechanical terms would demand mutually exclusive experimental arrangements.

Now analytical biology, from the time of Democritus onwards, has always been faced by the arguments of those who have seen in it only thanatology, or the study of death. Among the innumerable representatives of this point of view, which might be called "obstructionist organicism," we may select at random from the patristic period a passage from Tertullian: "The physician, or rather butcher, Herophilus of Alexandria, dissected six hundred persons that he might scrutinise nature; he hated man that he might gain knowledge. I know not whether he explored clearly all the internal parts of man, for death changes them from their state when alive, and death in his hands was not simply death, but led to error from the very process of cutting up."[37] Or, to come to our own time, the biologist Johnstone could say "Of living substance we literally know nothing. We study the behaviour only of the living organism. Whenever we study organic substance, it is necessarily dead inert material that we investigate."[38]

37. L. Septimius Tertullianus. *Opera* C. Guillard Paris 1545 fol. 250.

38. Johnstone, J., *The Mechanism of Life* (1921). Cf. also Eu-

Yet such statements can only be made by those who know nothing of biochemistry and experimental biology. With the advance of refinement in experimental technique, the injury done to the experimental material becomes ever slighter. And conversely we find that substantial parts of the normal set of properties of the intact living system remain operative when it is taken to pieces. As instances of the former proposition we may mention the technique of micro-manipulation, micro-dissection, and micro-injection,[39] or such work as the determination by spectrophotometric means of the pH in the cell-interior by the use of a naturally-occurring pigment,[40] or of the functions of cytochrome and the *Atmungsferment*.[41] Again, the use of X-ray analysis (of which more will be said in the third of these lectures) permits of far-reaching conclusions about the struc-

genio Rignano: "L'Analyse chimique ne peut se faire que sur la substance organique *morte*," *Scientia* (1929), 124; and Henri Bergson: "It is only with facts of katagenic order that physicochemistry deals—that is, in short, with the dead and not with the living." *Creative Evolution* (London, Macmillan, 1913), p. 37.

Thanatological dogma can also be found in its purest form in the writings of E. W. MacBride. Occasionally it deceives even the elect (cf. *Biological Principles*, *loc. cit.*, pp. 288, 294).

39. Cf. Chambers, R., "The Nature of the Living Cell as Revealed by Micromanipulation," contribution to *Colloid Chemistry Theoretical and Applied*, ed. J. Alexander (New York, 1928), vol. ii.

40. Vlès, F., and Vellinger, E., "Recherches sur le pigment de l'oeuf d'*Arbacia* envisagé comme indicateur de pH intracellulaire," *Archives de physique biol.*, *6* (1928), 239.

41. Keilin, D., "Cytochrome and Intracellular Respiratory Enzymes," *Ergebn. d. Enzymforsch.*, *2* (1933), 239; Warburg, O., many papers in *Biochem. Zeitschr.*, 1918 onwards.

ture of the protein framework.[42] Or valuable information about the metabolism can be drawn from a study of the effects of temperature upon the rhythmical or irregular activities which they exhibit,[43] as may be seen from work which causes no greater inconvenience to the experimental protozoa than a swim through a glass tube under the microscope at different temperatures within the limits of normal life.[44] Consider again what a vast system of knowledge concerning the mechanism of heredity and evolution has been gained simply by the controlled crossing of plants and animals, and the comparison of this evidence with the results of examination of dead fixed microscopic specimens. Even in cytology, the technique of tissue culture permits of the study of cells removed from the influence of the body as a whole. Then conversely, a wide array of respiratory and metabolic phenomena are found to occur normally in isolation, either by intact cells taken from the body, or even to a large extent in cell-free extracts, much of the organisation being destroyed.[45] But it

42. Sponsler, O. L., "Structural Units of Starch Determined by the X-ray Crystal Structure Method," etc., *Journ. Gen. Physiol.*, *5* (1923), 757; *9* (1926), 221, 677.

43. Crozier, W. J., "Distribution of Temperature Characteristics, and Critical Temperatures," *Journ. Gen. Physiol.*, *9* (1926), 525, 531.

44. Glaser, O., "Temperature and Mechanism of Locomotion in *Paramoecium*," *Journ. Gen. Physiol.*, *9* (1926), 115.

45. *See* p. 126. Sir F. G. Hopkins has a valuable passage on this point: "When living material is known to be uniform, so that the condition of a sample at any moment represents the condition of the whole, it is easy while a given reaction is proceeding to re-

would be tedious to run through the greater part of modern biology as evidence for what is already abundantly clear.

But Bohr proceeds to his second analogy. "The conditions holding for biological and physical researches," he says, "are not directly comparable, since the necessity of keeping the object of investigation alive imposes a restriction on the former which finds no counterpart in the latter. We should doubtless kill an animal if we tried to carry the investigation of its organs so far that we could describe the rôle played by single atoms in vital functions. In every experiment on living organisms there must remain an uncertainty as regards the physical conditions to which they are subjected, and the idea suggests itself that the minimal freedom we must allow the organism in this respect is just large enough to permit it, so to say, to hide its ultimate secrets from us. The asserted impossibility of a physical or chemical explanation of the function peculiar to life would in this sense be analogous to the insufficiency of the mechanical analysis for the understanding of the stability of atoms." Now the only answer to an anal-

move such samples at successive points along a time-scale. Progress of change in each successive sample can be instantaneously arrested, and estimations of this or that product duly made. The velocity curves of reactions may thus be obtained, though they have actually occurred in the living tissue. Studies of this kind supplemented by equally quantitative studies in tissue extracts, as well as other methods, thus make it possible to follow dynamic events. Only obscurantist thought, I think, will suggest that the data so obtained are unreal or incapable of application to the living cell itself." Purser Lecture, Dublin, *Irish Journ. Med. Sci.*, 1932.

ogy such as this is the denial that it has any force. It is not put forward with any single argument, and as its effect, if admitted, would be to sterilise investigation, it must be rejected. The misconceptions which lie at the back of it need not here be elaborated, but it displays a notably naïve view of the nature of death. There is one death when the metazoan body ceases to perform its normal functions, there is another death when the tissue-slice isolated from it ceases to glycolyse or to respire in the manometer, there is a third death when the cell-free enzyme preparation isolated from the tissue-slice ceases to catalyse its appropriate reaction. The assumption in the second sentence of the paragraph from Bohr just quoted has nothing to justify it.

It remains, therefore, to be shown that the nature of the organisation in the living system is the same as that in the atom, and it is not proven that the principle of uncertainty or indeterminacy in physics has anything whatever to do with the alleged thanatological limitation of biological theory.[46]

In history Meyer has little difficulty in illustrating his thesis[47] that mechanism follows at the heels of

46. Some time after I had written the preceding criticism of Niels Bohr, I found that the distinguished physiologist, Otto Meyerhof, had also felt the necessity of repudiating the analogy between the principle of indeterminacy and the difficulties of biological research. In defending the delicacy and validity of biological technique, he adds, to the instances given above, the work on the metabolism of populations of cells such as yeast, and the value of the *reversible* inhibitors now so commonly used for investigating the processes of intracellular chemistry. *Naturwiss, 22* (1934), 311.

47. *Ideen, loc. cit.*, p. 27.

mathematics like a dark shadow. Galileo's "new science of ancient things" with its aim to "measure all things and to make measurable what cannot yet be measured" was translated into biology by Harvey with his mathematical argument about the circulation of the blood,[48] and in lesser-known calculations, such as that of Freind[49] on the menstrual fluid, and Hartsoeker[50] on the *Emboîtement* of the preformationists; above all, by Borelli[51] in his work on muscular motion. But the lesser men could not resist taking the tempting step of assuming that all the processes of the body were readily to be explained on mechanical principles. Hence the astonishingly crude speculations of the Iatro-mathematical school, represented, e.g., by Baglivi and Scarborough; and I fear we must include among the lesser men so far as biology is concerned the great Cartesius himself,[52] and certainly Gassendi. It is only by reaction from their successors that Haldane can write: "We are unable to apply mathematical reasoning to life, since mathematical treatment assumes a separability of events in

48. Harvey, W., *De Motu Cordis* (Frankfurt, 1628), chap. ix.
49. Freind, J., *Emmenologia* (London, Cox, 1729).
50. Hartsoeker, N., *Essai de Dioptrique* (Paris, 1694).
51. Borelli, G. A., *De Motu Animalium* (Rome, 1680).
52. A perfectly representative sample can be taken from the *De Homine Liber,* 1662, on the motion of the heart. "Then, because the little parts now dilated tend to continue their movement in a straight line, and because the heart, now newly formed, resists them, they move away from it and take their course towards the place where afterwards the base of the brain will be formed. Thus they enter into the place of those that were there before, which for their part move in a circular route to the heart, and there, after waiting for a moment to assemble themselves, dilate and follow the same road as the aforementioned ones."

space which does not exist for life as such. We are dealing with an indivisible whole when we are dealing with a life."[53] To close the door in this way to the possibilities of the integral calculus, Topology and *Mengenlehre* applied to biology is a gratuitous denial of the intellect.

It is more pleasant to turn to glance at some of the pioneer investigations in the mathematisation of morphology. Outstanding, of course, is the work of D'Arcy Thompson, summarised in his classical *Growth and Form*.[54] In that remarkable book, he applied mathematical, especially geometrical, analysis to large tracts of morphology, treating of the internal form and structure of the cell, forms of cells and tissues, concretions, spicules, and skeletons, the logarithmic spiral, the shapes of horns, shells, leaves, and eggs, and the relation between form and mechanical efficiency. One instance only of the kind of work which he carried out may be given from his theory of transformations in which he showed that a geometrical relation exists between related animal forms. The process of comparison, of recognising in one form a definite permutation or deformation of another, apart altogether from a precise and adequate understanding of the original "type" or standard of comparison, lies within the immediate province of mathematics, as D'Arcy Thompson said, and finds

53. Haldane, J. S., *loc. cit.*, p. 16. *See* the comments of Meyer on this passage in "Die Axiome der Biologie," *Nova Acta Leopoldina*, N.F. *1* (1934), 474, p. 481 ff.

54. Thompson, D'Arcy W., *Growth and Form* (Cambridge, University Press, 1917), *see especially* chap. xvii.

its solution in the elementary use of the method of co-ordinates. In a net of rectangular co-ordinates it is possible to inscribe the outline of a fish or other animal, and so to translate it into a table of numbers. However complicated its outline, this will be, in general terms, a function of x and y. If the rectangular system is now submitted to deformation by altering the direction of the axes from a right angle to some other angle (altering the ratio x/y), or substituting for x and y some more complicated expressions, then a new set of co-ordinates is obtained, whose deformation from the original type the inscribed figure will precisely follow. The new figure represents the old figure under strain, and is a function of the new co-ordinates in exactly the same way as the old one was of the original co-ordinates.

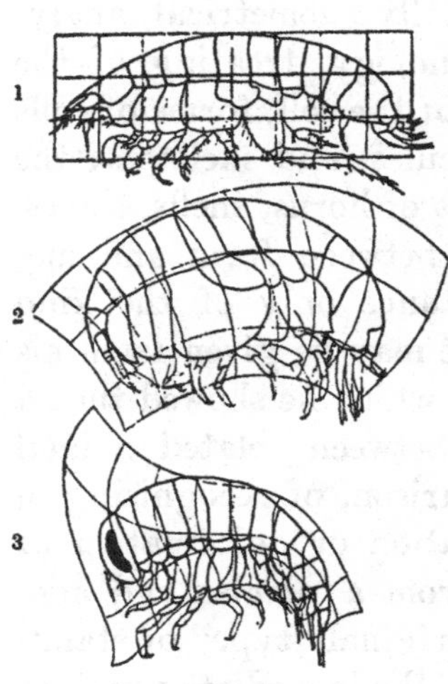

FIG. 1.

Related species of Amphipods described in Thompsonian co-ordinates.
1. *Harpinia plumosa* Kr.
2. *Stegocephalus inflatus* Kr.
3. *Hyperia galba.*

Let us take as an example, though it be a complicated one, the result of applying this method to the generic and specific variations found in the crustacean group of amphipods (Fig. 1). In related organisms one particular mode of variation is usually so prominent and so paramount throughout the entire organism that one comprehensive system of co-ordinates suffices to give a fair picture of the actual phenomenon. Deforming the rectangular co-ordinates of *Harpinia* into a curved orthogonal system, a very fair representation of an allied genus, *Stegocephalus*, is obtained. With greater deformation of the co-ordinates, a tolerable representation of the aberrant genus *Hyperia* appears, with its narrow abdomen, reduced pleural lappets, great eyes, and inflated head. Another example may be taken from the fishes (Fig. 2). Here *Diodon*, the porcupine-fish, is plotted on rectangular co-ordinates, and, in the adjacent figure, these have been transformed into concentric circles vertically and hyper-

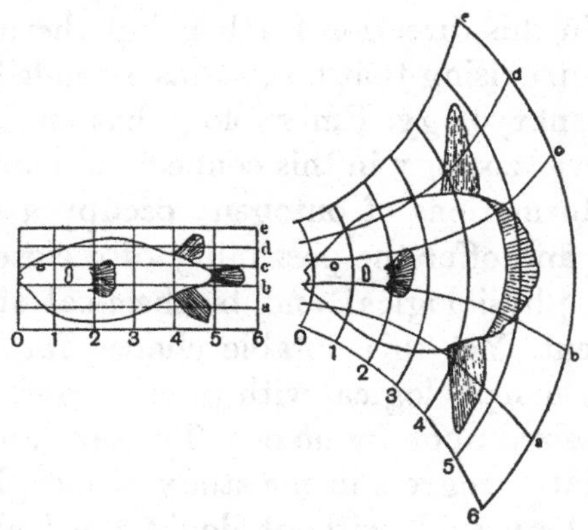

FIG. 2.

Related species of Fishes described in Thompsonian co-ordinates.
Left. *Diodon*.
Right. *Orthagoriscus*.

bolas horizontally. The old outline, transferred to the new network, appears as a clear representation of the closely allied sunfish, *Orthagoriscus*. Here the mathematical transformation is not regular, as the system is no longer isogonal, but it is nevertheless symmetrical to the eye and obviously approaches an isogonal system under conditions of friction or constraint. And as such it accounts by one single fundamental transformation for all the apparently separate and distinct external differences between the two fishes.

D'Arcy Thompson's main interest in these studies was evolutionary or phylogenetic, as was shown by his brilliant treatment of geometrical transformations in the series of skulls of mammals (e.g., *Hyracotherium*, *Mesohippus*, *Protohippus*, etc.). It is astonishing that he has had so few successors to carry his work in this direction further, but therefore perhaps not surprising that no one has extended it to the realm of embryology. Embryology has one great advantage over zoology in this connection, namely, that the transformations of ontogeny occupy a very limited time, and offer the possibility of a close correlation with physiological and biochemical studies on development. Yet this notable chance for the comparison of morphological with physico-chemical data has been seized upon by no one. The severest obstacle to immediate progress in the study of morphogenesis is, it has been said, without doubt the lack of any quantitative measure of differentiation. Physico-chemical changes can to-day be correlated with growth-rate, as in much work on susceptibility to

X-ray injury,[55] but no parallel correlations with differentiation are as yet possible. It can easily be seen how difficult the problem is, for any formulation of it would have to include parameters for the number of organs (cellular parts), the number of cells, the number of types of cells or organs, the number of active physiological functions, the areas of folded or puckered surfaces, the number of tubes or branches of tubes between given points, etc. D'Arcy Thompson's method of transformations, applied to a standard number of cross-sections of the developing embryo in all three dimensions at regular intervals of time, might be a beginning worth making.

So far, only one attempt to measure differentiation-rate has ever been made, that of Murray.[56] All who have had any practical contact with embryology know that there is much more difference in shape and form between a chick embryo of the second day and one of the fifth than there is between one of the twelfth day and one of the fifteenth. Murray made an admirably ingenious attempt to evolve a method of evaluating change in shape quantitatively, and chose as the organ for investigation the heart, which, as far as weight is concerned, appears to keep pace exactly with the embryo as a whole. This was tested also by measurements of the surface area of the

55. Cf. the very quantitative paper of Voskressensky, N. M., "Uber d. Wirkung d. Röntgenstrahlen auf das embryonale Wachstum," *Archiv. f. Entwicklungsmech.*, *113* (1928), 447.

56. Murray, H. A., "Physiological Ontogeny, VIII, Accelerations of Integration and Differentiation during the Embryonic Period," *Journ. Gen. Physiol.*, *9* (1926), 603.

organ, calculating the percentage growth-rate with the aid of a projectoscope and a planimeter. He next resorted to the expedient of selecting forms spaced by a visual impression so as to represent approximately equal degrees of gross change. In other words,

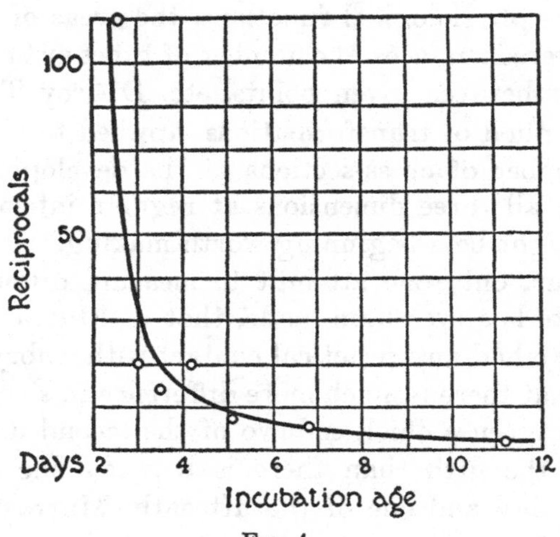

FIG. 4.

Attempt at a quantitative formulation of differentiation-rate. (From Murray.)

from a series of drawings made at frequent intervals, certain ones were chosen which seemed by inspection to be equally spaced from one another in respect to their relative complexity of form. The test was thus necessarily arbitrary and open to criticism because of its subjective nature. By taking the average of many embryos it was then determined what were the

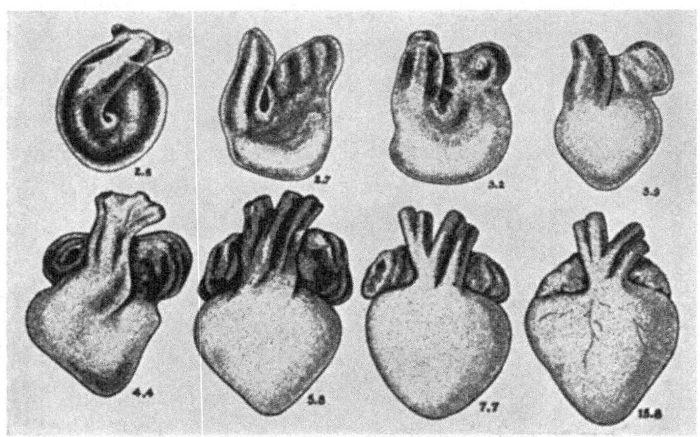

Fig. 3.

Drawings of the development of the heart of the chick embryo. (From Murray.) The numbers indicate the day of incubation.

exact incubation ages of the embryos with heart-forms such as those selected. The illustration (Fig. 3) shows these clearly. The reciprocals of the time intervals between successive drawings were then used as rough criteria of the rate of form development. The graph resulting from the plotting of these against time is shown in Fig. 4, from which it is clear that the rate of evolution of morphological form falls precipitously at first and then ever more slowly, essentially resembling in this way the instantaneous percentage growth-rate. Growth-rate and differentiation-rate seem thus to be aligned with one another, for the most marked form-changes occur at the initiation of embryonic development when the growth-rate is at its highest. The shifts of chemical equilibrium in the embryo, it is interesting to note, follow a rather different course, but that is another story which cannot be given here.[57]

In these attempts toward the mathematisation of morphological form we find ourselves on the threshold of solution of the problem of the relative rates of growth and differentiation, a problem which was clearly formulated by, and greatly perplexed, Aristotle,[58] and which for centuries lost itself, by way of Talmudic and Patristic speculation on "perfection-time," in the arid regions of semi-theological speculation.[59]

57. Consult Needham, J., *Chemical Embryology* (Cambridge, University Press, 1931), I, 547 ff.

58. *De Generatione Animalium,* §§ 743a, 743b.

59. *See* Needham, J., *A History of Embryology* (Cambridge, University Press, 1934).

The striking dissociability of differentiation and growth is another fact which should lead us to doubt the intuition of those who emphasise the unanalysability of the organism. In normal development of an animal embryo the fundamental processes are seen as constituting a perfectly integrated whole. They fit in with each other in such a way that the final product comes into being by a precise co-operation of reactions and events. But by means of various experimental procedures these processes can be dissociated or thrown out of gear with one another. Thus embryonic growth can be stopped without abolishing embryonic respiration, and conversely, it is probable that growth or differentiation, under certain conditions, may proceed in the absence of normal respiratory processes. There are many instances, too, where growth and differentiation are quite separable.[60] But it would lead too far to follow further this interesting train of thought.

A few words more may be devoted to the characteristics of vitalism as found in the writings of Haldane. Its self-contradictory tendency is well illustrated in his Donnellan lectures, previously referred to. On one page[61] we are told that the central facts relating to the specificity of a species are part of the residuum which physico-chemical biology is powerless to touch. Yet a few pages further on we read

60. *See* Needham, J., "On the Dissociability of the Fundamental Processes in Ontogenesis," *Biol. Rev., 8* (1933), 180.
61. Haldane, J. S., *loc. cit.*, p. 11.

"the constitution of proteins, including haemoglobin, which can be separated from the bodies of different individual men, varies appreciably between different individuals; and this is no mere accidental circumstance, but is as characteristic for the individual as the shape or size of his hands or face, or the colour of his hair."[62] Two thinkers struggle perpetually for the possession of Haldane, one the exact biologist, full of desire for the metrical description of phenomena and for their causal explanation, the other the idealist philosopher, primarily concerned with making the world safe for the spirit, and not averse from frequent appeals to the alogical core of the universe. It is for failure to differentiate between the two constituents of this dual personality that successive generations of biologists have been vexed and irritated by the counsels of one of the greatest among them.

I do not wish in the course of this train of thought to embark upon any discussion of the place which psychology should hold in our estimate of biology, but it is difficult to avoid commenting upon the vitalist's use of the word "essential." "Consciousness," writes Haldane, "can certainly be treated from a purely mechanistic standpoint if we pay no attention to essential facts, just as other sides of physiology can be so treated when essential facts are left out of account."[63] The only thing lacking here is a definition of the word "essential." There is no logical dif-

62. *Idem,* p. 22.

63. *Idem,* p. 87.

ference between this position and that of a theologian who would object to the discussion of consciousness without reference to the facts of religious experience. We may not like sometimes to use the method of abstraction, but it is an unavoidable part of scientific methodology. This leads to an underestimate of the work of Pavlov and his school, as several students of Haldane's writings, notably Hogben, have often remarked.[64] To make use of an Eddingtonian phrase, when Haldane says "mind" he means "mind (loud and prolonged cheers)," that is "mind" in the sense of Berkeley and Bosanquet, not that of Freud and Pavlov. Haldane is assured of an honourable place in the history of biology in the late nineteenth and early twentieth centuries. But this will always be in spite of, and not because of, his presentation of his case. It will be said of him that he took organisation seriously when no other biologist would, but perhaps this recalcitrance on the part of his colleagues was partly due to his own inability to distinguish between the conflicting claims of science and philosophy, and even, we may add, of religion also.

It is refreshing to turn to the writings of a very different school, namely, the dialectical materialists. Although this philosophy has been widely applied in recent years to biological problems, the great divisions of language and culture which separate the U.S.S.R. from the European West have greatly curtailed its influence. For this reason the contributions

64. Hogben, L., *The Nature of Living Matter* (London, Kegan Paul, 1930), p. 33 ff.

of Zavadovsky[65] and Colman[66] to the Congress of the History of Science in 1931 were of particular interest, and since then there have been interesting publications by Hecker,[67] Prenant,[68] and Schaxel.[69]

In the short space that remains it would obviously be impossible to attempt to do justice to the general world-picture of dialectical materialism. What I wish to point out here is that, in biology, the standpoint which follows from its fundamental propositions is closely similar to that which I have been supporting in this lecture. In the proper definition of vitalism, in the misconceptions surrounding the use of the words "axiomatic" and "irreducible," in the attempt to equate the uncertainty principle of physics with the thanatological difficulty in biology, we have repeatedly seen that biological order is a form of order different from those found in physics, chemistry, or crystallography, yet not impenetrable by the human mind or ruled by unintelligible spiritual entities. Translated into terms of Marxist philosophy, it is a new dialectical level.

65. Zavadovsky, B., "The 'Physical' and 'Biological' in the Process of Organic Evolution," contribution to *Science at the Cross-roads* (London, Kniga, 1931).

66. Colman, E., "Dynamic and Statistical Regularity in Physics and Biology," contribution to *Science at the Cross-roads* (London, Kniga, 1931).

67. Hecker, J., *Moscow Dialogues* (London, Chapman & Hall, 1933), *see especially* pp. 161, 162.

68. Prenant, M., "La conception matérialiste dialectique en Biologie," *Bull. Soc. Philomath., 116* (Paris, 1933), 84.

69. Schaxel, J., "Das biologische Individuum," *Erkenntnis, 1* (1930), 467. Also "Leninismus und Biologie" article in *АКАДЕМИЯ НАУК СССР ПАМЯТИ В. И. ЛЕНИНА* 1924–1934, Moscow (German translation).

"Dialectical materialism," says Prenant, "is as much opposed to vitalism as to old-fashioned mechanistic materialism, because both are metaphysical theories. It refuses to trace any sharp demarcation between physics and biology, reserving to the former causal determinism, and appealing in the latter to teleology. But it does not believe, on the other hand, that biology has the task of reducing itself wholly and effectively to physics. The unity of the universe expresses itself in qualitatively different forms, the characters proper to which must not be lost sight of." Or, in Zavadovsky's words: "The true task of scientific research is not the violent identification of the biological and the physical, but the discovery of the qualitatively specific controlling principles which characterise the principal features of every given phenomenon, and the finding of methods of research appropriate to the phenomenon studied. . . . Affirming the unity of the universe and the qualitative multiformity of its expression in different forms of motion of matter, it is necessary to renounce both the simplified reduction of some sciences to others, and the sharp demarcation between the physical, biological, and socio-historical sciences." All this is a repetition in different terms of the principle that biological order is both comprehensible and different from inorganic order.

Dialectical materialism, indeed, has two aspects which are especially relevant here. As Bernal[70] puts

70. J. D. Bernal's essay in *Aspects of Dialectical Materialism*, by H. Levy, J. Macmurray, R. Fox, R. P. Arnot, J. D. Bernal,

it, the central idea of this philosophy is that of transformation. How do transformations occur, and how can we make them occur? Any satisfactory answer must also be a solution of the problem of the *origin of the qualitatively new.* I shall return to this point toward the end of the last lecture.

It would, I believe, be easy to show that the standpoint of Dialectical Materialism on these questions has been held in one way or another by numerous interesting precursors, probably of all countries, and certainly in the English-speaking ones. From the period of romantic biology and medicine one could choose at once S. T. Coleridge[71] and James Hinton.[72] The former, in his "Essay Towards the Formation of a More Comprehensive Theory of Life," declared that life was a process, or mode of operation, of the same powers which we recognise under other names, such as magnetism, electricity, or chemical affinity. These, by their own properties, effected all the results observed in life, but were grouped in a special way, the various forms of action being so united as to constitute, out of many parts, a mutually dependent whole. James Hinton, in his *Life in Nature* (1862) adopted the same attitude. "The organic world," he wrote, "does not differ from the inorganic in its es-

and E. F. Carritt (London, Watts, 1934). See also "Engels and Science," *Labour Monthly 17* (1935), 506.

71. Coleridge, S. T., "Essay Towards the Formation of a More Comprehensive Theory of Life." *Miscellanies, Aesthetic and Literary* (London, Bohn, 1885).

72. Hinton, James, *Life in Nature,* first published 1862, republished with introduction by Havelock Ellis (London, Allen & Unwin, 1932).

sence. But it differs. It would be a fatal error—happily it is an impossible one—to confound the two. There is a difference in the mode of operation, though the elements are the same. The physical powers have received in the organic world a particular direction, and are made to work to certain results which are attainable only through living structures." And Hinton, criticising the vitalists, says: "For the peculiar results, a peculiar agent was supposed, instead of a peculiar mode of operation."

To sum up, then, we have considered in this lecture some of the opinions which biologists, physicists, and philosophers hold regarding the form of organisation which living things exhibit. To discuss biology without having in mind the existence of the form-problem, is, as we have seen, to wander on the outer fringes of the science. On the other hand, to discuss it with biological organisation regarded as something fundamentally inscrutable is at least equally futile.

II

THE DEPLOYMENT OF BIOLOGICAL ORDER

IN the preceding lecture the discussion of biological organisation was somewhat abstract. We did not consider the fact that it is present more obviously in some situations than in others, in the adult organism, for instance, more obviously than in the egg. Such an increase in "visibility" of order as occurs during animal development we may speak of as "deployment."

In pursuing this theme, we may begin by giving some account of the contributions of Hans Driesch to the philosophy of morphogenesis. Although his theory of the entelechy is now largely a matter of historical interest, the investigations which led to it were quite fundamental. We are so accustomed nowadays in embryology to the concept of determination, i.e., the fixing of the fates of parts of the embryo at a definite time in development, that it is difficult to think back into the state of mind in the last century when it was supposed that fates were all fixed to begin with.

In 1883, about the same time as Weismann, W. Roux[1] enunciated the view that development was brought about by a qualitative division of the germplasm contained in the nucleus, and that the compli-

1. Roux, W., *Über die Bedeutung d. Kernteilungsfiguren* (Leipzig, 1883).

cated process of mitotic division was primarily devoted to that end. It was thought that development proceeded by a mosaic-like distribution of potencies to the cells in segmentation, and that for instance the

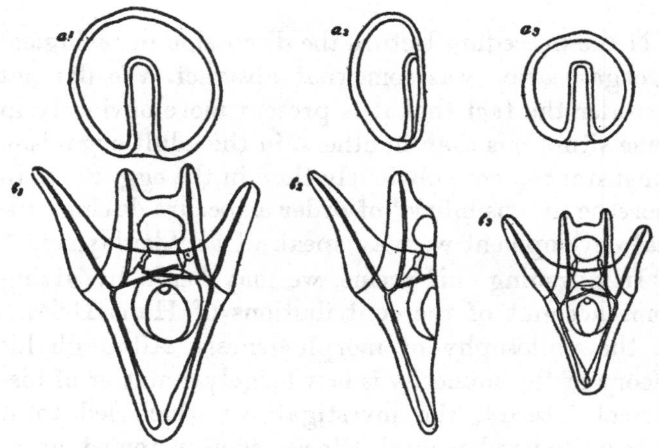

FIG. 5.

Experiments on pluripotence in *Echinus*. (From Driesch.)
a_1 and b_1. Normal gastrula and normal pluteus.
a_2 and b_2. Half-gastrula and half-pluteus, expected by Driesch.
a_3 and b_3. The small but whole gastrula and pluteus, which he actually obtained.

first cleavage-furrow separated the material and the potencies of the right side from those of the left. In 1888 Roux announced[2] that, if one of the two first blastomeres of a frog egg was killed by cautery with a hot needle, a half-embryo developed from the unin-

2. Roux, W., *Archiv. f. Pathol. u. path. Anat.* (Virchow's), *114* (1888), 113.

jured cell. This "mosaic" development seemed to confirm the view of Weismann[3] that the fates of all the parts were fixed at the outset, and, although Roux's particular experiment afterwards turned out to be deceptive, in that the amphibian blastomeres are not necessarily mosaic, a large number of other eggs were subsequently found in which any injury to the uncleaved or cleaving ovum is reflected in injuries or losses in the finished embryo.[4]

Considerable astonishment was therefore caused by Driesch's announcement in 1891[5] that he had obtained complete larvæ from single blastomeres of the sea-urchin egg isolated at the two-cell stage. This is illustrated in the picture with which he afterwards furnished his Gifford Lectures[6] (Fig. 5), and we may give his account of the discovery in his own words: "The development of our *Echinus* proceeds rather rapidly, the cleavage being complete in about fifteen hours. I quickly noticed on the evening of the first day of the experiment, when the half-embryo was composed of about two hundred cells, that the margin of the hemispherical embryo bent together a little, as if it were about to form a whole sphere of smaller size, and indeed the next morning there was a whole di-

3. Weismann, A., *Die Kontinuität des Keimplasmas als Grundlage d. Vererbung* (Jena, 1885).

4. For a fuller account, *see* E. S. Russell's admirable history of nineteenth-century biology, *Form and Function* (London, Murray, 1916).

5. Driesch, H., *Zeitschr. f. wiss. Zool., 53* (1891), 160.

6. Driesch, H., *The Science and Philosophy of the Organism*, Gifford Lectures (1st ed. 1908; 2d ed., London, Black, 1929). The references here are to the latter issue.

minutive blastula swimming about. I was so much convinced that I should get the Roux effect in all its features that, even in spite of this whole blastula, I now expected that the next morning would reveal to me the half-organisation of the subject once more; the intestine, I supposed, would come out on one side

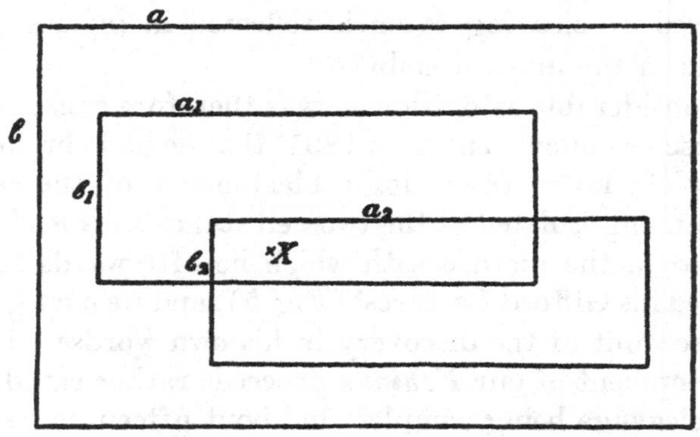

FIG. 6.

Driesch's diagram to show the characteristics of a harmonious equipotential system. The element X forms part of the systems a,b, or a_1b_1, or a_2b_2; its prospective value is different in each case.

of it, as a half-tube, and the mesodermal ring might be a half one also. But things turned out as they were bound to do, and not as I had expected; there was a typically *whole* gastrula in my dish the next morning, differing from a normal one only by its smaller size, and this small but whole gastrula was followed by a whole and typical small pluteus larva."

In the following year he showed that whole embryos could be produced from one or more blastomeres isolated at the four-cell stage,[7] and later that alteration of cleavage by compression,[8] the fusion of two eggs into one,[9] or the cutting in certain directions of the original egg-cell,[10] all permitted of normal embryos being produced.

Driesch introduced the term "prospective significance" (*prospektiv Bedeutung*) to indicate the actual fate of any part or monad in the original egg. His great discovery lay in the finding that the significance of such a part was not exhausted by its prospective significance, but that it was widely changeable according to various circumstances. Next he introduced the term "prospective potency" (*prospektiv Potenz*) in order to signify the collection of possible fates of such a part. Thus the actual fate (the prospective significance) is chosen from among the possible fates (the prospective potency); in nor-

7. Driesch, H., *Zeitschr. f. wiss. Zool., 55* (1893), 1. "Die isolirten Blastomeren d. Echinidenkeimes," *Archiv. f. Entwicklungsmech., 10* (1900), 361.

8. Driesch, H., *Mitt. a. d. zool. Sta. Neapel., 11* (1895). "Betrachtungen ü. d. organisation d. Eies u. ihrer Genese," *Archiv. f. Entwicklungsmech., 4* (1897), 75, *especially* p. 113; "Neue Ergänzungen zur Entwicklungsphysiologie d. Echinidenkeimes," *Archiv. f. Entwicklungsmech., 14* (1902), 500, *especially* p. 517.

9. Driesch, H., *Zeitschr. f. wiss. Zool., 55* (1893), 1. "Verschmelzung d. Individualität bei Echinidenkeimen," *Archiv. f. Entwicklungsmech., 10* (1900), 411.

10. Driesch, H., "Zur Analyse der Potenzen embryonaler Organzellen," *Archiv. f. Entwicklungsmech., 2* (1895), 169; "Drei Aphorismen zur Entwicklungsphysiologie jüngster Stadien," *Archiv. f. Entwicklungsmech., 17* (1903), 41; "Altes u. Neues zur Entwicklungsphysiologie d. jungen Asteridenkeimes," *Archiv. f. Entwicklungsmech., 20* (1905), 1.

mal development it will be one constant result, in abnormal development it may be quite a different one. This condition of multiple potency of the parts of the early egg-cell has been termed Pluripotence.

These relationships were illustrated by Driesch[11] in the accompanying diagram (Fig. 6). The plane of the dimensions a and b is the basis of the normal undisturbed development; taking the sides of the plane as fixed localities for orientation, we can say that the actual fate (the prospective significance) of every element of the plane stands in a fixed and definite relation to the length of the two co-ordinates at right angles. Thus the point X would normally develop into some definite part of the finished embryo, e.g., eye-cup or muscle-cell. But its prospective significance is not as wide as its prospective potency, so that if it formed part of an isolated portion a_1b_1 its relation to the whole would be different and its actual fate would be different. The same applies to its position in another possible isolated portion a_2b_2. These isolated portions, it is to be remembered, regulate themselves in such a way as to reproduce on a smaller scale the relative location of parts seen in the absolutely normal case. They possess, therefore a nonvariable factor of wholeness, represented in Driesch's terminology by the symbol E.[12] In view of these properties, systems which fulfilled the conditions just laid down for the plane ab were called by Driesch "harmonious equipotential systems."

11. Driesch, H., Gifford Lectures, p. 90.
12. *Idem,* p. 91.

We know now, after forty years of experimental research,[13] that at the beginning of development, parts or monads of the egg, similar to that which has been represented by X in the diagram, are indeed undetermined,[14] and that one of the most fundamental processes in development consists in the closing of doors, i.e., in determination, in the progressive restriction of the possible fates.

By the method of transplantation evolved by Spemann, it has been shown, in the case of the newt, for example, that up to a certain stage of gastrulation, the fates of most of the embryonic regions are not irrevocably determined. A piece of presumptive neural tube material removed from one embryo and grafted into another will turn into external gills if it happens to be grafted into the presumptive gill-region of the latter. On the other hand, a piece of presumptive skin, if grafted into a suitable region of the presumptive nerve-tube of a second embryo, will in due course turn into brain or spinal cord. Up to this stage in gastrulation, therefore, the regions develop always in step with their actual surroundings and without reference to their former surroundings,

13. Summarised in Huxley, J. S., and de Beer, G. R., *Elements of Experimental Embryology* (Cambridge, University Press, 1934).

14. That is, undetermined embryologically, not genetically. The sense in which the word "determination" is here used must be distinguished from other usages; 1) the "determination" of the original axes of the egg by external factors (Roux); 2) the "determinate cleavage" of eggs in which the cleavage pattern is very definite and permits of accurate descriptions and predictions of cell-lineage (Conklin); 3) "determinants," hypothetical units of the germ plasm (Weismann).

or their prospective significance if they had not been interfered with. They are absolutely plastic, like the parts in Driesch's sea-urchin eggs, which would obediently form gut or skeleton according to their position in the whole. Nor does this plasticity or pluripotence stop short at the germ-layers themselves, for mesoderm and ectoderm, for example, are perfectly interchangeable.

The process of gastrulation, however, is critical, for at this time the original plasticity is lost, and the main fates of the parts are determined. This is irrevocable. The early period of plasticity is followed by one of rigidity, each part being willing now only to undergo a certain special type of development which differs from part to part. If it is grafted into another embryo, it will continue to differentiate in accordance with its inner determination, and not in accordance with the new situation in which it has been artificially placed. The regions can now only develop toward their presumptive fates. Prospective potency has been ruthlessly curtailed to prospective significance. The eye-region will form an eye whether this points outward to the external world or inward to the body-cavity. The invisible process of determination has ushered in the new period of self-differentiation, in which the embryo has become a mosaic of irreplaceable regions, similar to the whole development of certain eggs ("mosaic-eggs"), which never manifest a period of plasticity or pluripotence.

One region of the amphibian embryo, however, is

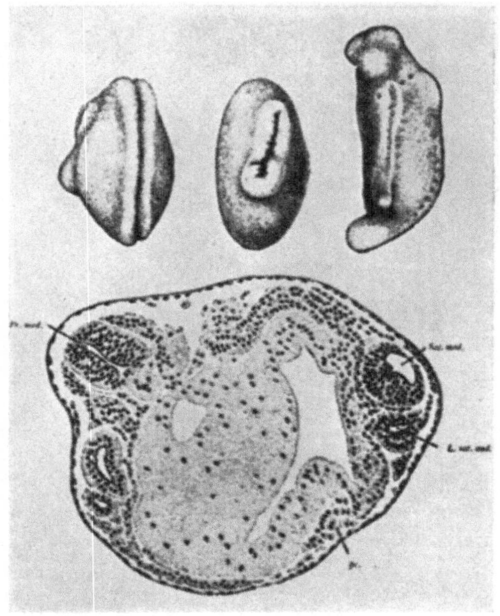

Fig. 7.

Action of a piece of organiser from an embryo of the newt *Triton cristatus* implanted into an embryo of the newt *Triton tæniatus*, showing the induction of a secondary embryo in the host. (From Spemann and Mangold, after Dürken.)

Top left. Dorsal view of *tæniatus* embryo, showing the host's neural groove and folds along their whole length; to the left on the side, the secondary neural folds. About 44 hours after implantation.

Top middle. Side-view of the same embryo, showing the secondary neural groove.

Top right. Side-view of the same embryo 5 days after the operation. The secondary embryo is seen from above, with its tail-bud, neural tube, somites, and auditory vesicle.

Below. Transverse section through the embryo at the level of the primary pronephros. The primary axial organs are above at the left, the secondary or induced ones are above at the right; Pr. med. primary neural tube; Sec. med. secondary neural tube; L. sec. aud. left secondary auditory vesicle; pc. pericardium.

not plastic during the early part of development. This is the region of the dorsal lip of the blastopore (Fig. 7), which has arisen from the grey crescent and will in time form the notochord and mesoderm. This is determined from very early stages, possibly before any cleavage has begun. When grafted into other embryos it will develop in no direction other than its usual presumptive fate. If a piece of this region is grafted into another embryo in the blastula or early gastrula stage, it will "induce" the neighbouring host tissues to form a secondary embryo, often including nerve-tube, brain, eyes, ears, somites, notochord, etc., irrespective of the presumptive fates of those tissues. In other words, it contains within it the influence which determines the fates of regions with which it comes in contact. It is therefore called the "organiser" (*Organisator* of Spemann). It is easy to demonstrate its fundamental importance in the development of the amphibian egg by separating the first two blastomeres instead of killing one of them, as in Roux's experiments. Their subsequent fate will now depend on whether or not they contain a sufficiency of the region which will afterwards become the organiser region. If both blastomeres contain a portion of it, both will be organised into small but morphologically normal embryos, in accordance with the original experiment of Driesch. If one contains all of it and the other contains none (as will happen if the first cleavage chances to be horizontal) the former will produce a complete embryo and the latter will give up the struggle, as it were, after a few cleavages. We now understand better the conditions necessary for the fulfilment of Driesch's pluripotence.

Development, then, consists of a progressive restriction of potencies by determination of parts to pursue fixed fates. It is the conviction of many that this state of affairs can best be pictured in the manner of a series of equilibrium states. The whole process could then, as Waddington suggests,[15] be imagined in the likeness of a series of cones (Fig. 8). At the top of the uppermost cone there is a ball in a position of extremely unstable equilibrium. It will tend to fall along the side of the cone and will reach a point at some one of the 360 degrees of the cone's circumference. Here it will again find itself in a position of unstable equilibrium, only now with respect to a second stage of determination, and will again be pushed

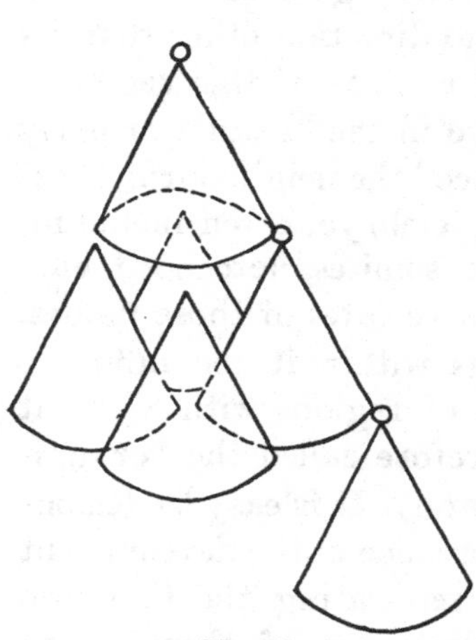

FIG. 8.

Waddington's cones; a diagrammatic way of representing the course of embryonic determination.

15. Waddington, C. H., "Experiments on the Development of Chick and Duck Embryos, cultivated in vitro," *Phil. Trans. Roy. Soc.*, *221* (1932 B), 179, p. 221.

in one direction or another, again to occupy a passing equilibrium, and so on until the final stage of absolute stability is reached, i.e., the plan of the adult body. In biological terms a piece of ectoderm will be determined first of all by the primary organiser to form head-ectoderm, next it will be determined by the secondary organiser of the eye-cup to form the lens of the eye, and so on. The effect of the primary and secondary organisers in the normal course of events is to give the slight pushes necessary to send the balls down the cones into their positions of greater stability as time goes on. A grafted organiser will give a push of this kind to a given ball (plastic region) in a direction contrary to that which it would normally have taken, but for this it is necessary that the ball should be high up the levels of instability, for if it has already been severely determined, the grafted organiser can have no effect upon it. Conversely, an organiser of the second or third grade, as it is called (Spemann[16]), has no effect upon a ball (plastic region) at a higher level of instability than that at which it normally works. The undetermined material is then said not to be "competent" to receive its action (Waddington[17]).

This concept of unstable equilibria in early development seems to offer opportunities for advance in the mathematisation of embryology. Lotka,[18] who has

16. Spemann, H., Croonian Lecture, *Proc. Roy. Soc., 102* (1927 B), 177.

17. Waddington, C. H., *loc. cit.*, p. 223.

18. Lotka, A. J., *Elements of Physical Biology* (Baltimore, Williams & Wilkins, 1925), pp. 77, 146, 294, etc.

devoted much study to the use of equilibrium-concepts in biology, gives a list of the various types of equilibria which may be met with. If only two variables are in question, the integral curves may be plotted on rectangular co-ordinates. Families of curves will be obtained passing through the equilibrium points, which will appear as singular points. As an example, Lotka figures the topographic chart of the well-known Ross malaria equations,[19] which describe the course of events in the spread of malaria in a human population by the bites of certain breeds of mosquitoes infected with the malaria parasite (Fig. 9). In this chart there are two singular points, one at the origin, O, unstable, the

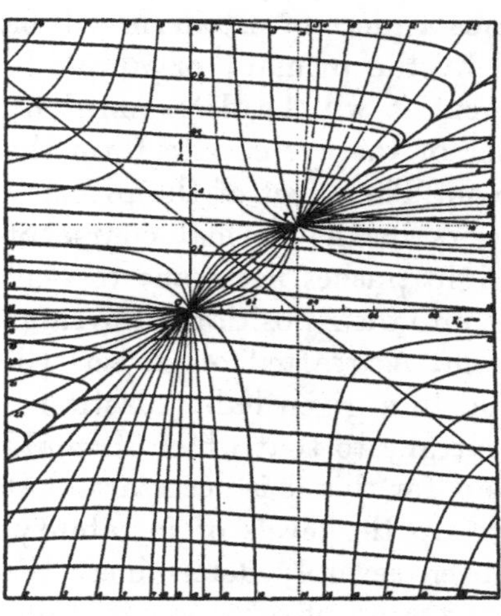

FIG. 9.

Map of integral curves for the Ross malaria equations. The heavy lines are integral curves; the lighter lines are auxiliaries (isoclines) employed in constructing the graphic solution of the differential equations; there are two singular points. (From Lotka.)

19. Ross, Sir R., *The Prevention of Malaria* (2d ed. 1911), p. 679; *also* Lotka, A. J., *Amer. Journ. Hygiene,* 1923. (Jan. Suppl.)

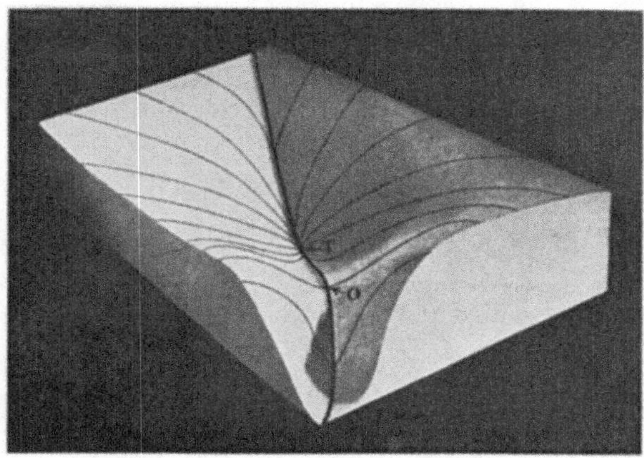

Fig. 10.

Model showing interpretation of the integral curves of Fig. 9 as lines of flow, or lines of steepest descent on a topographic surface. The pit at the centre of the model corresponds to the point of stable equilibrium (T in Fig. 9); the centre of the notch (the point of intersection of the diagonal curve and the second integral curve counting from the front corner) corresponds to the point of unstable equilibrium (O in Fig. 9). The correspondence between these figures is qualitative only. (From Lotka.)

Fig. 11.

Qualitative three-dimensional model of embryonic determination, illustrating the passage from unstable to stable equilibria.

other at T, stable. The chart, Lotka points out, obviously suggests stream-lines and a three-dimensional model. Such a model is shown in Fig. 10. The feature of interest is that a singular point like O is represented by a col or saddle-point in the landscape, so that the attainment of stable equilibrium would be represented by the fall of a point from O to the pit T. The model serves, as Lotka says, to bring out an important fact, namely, that there are necessarily certain regularities in the occurrence of the various types of equilibria. Thus two pits of the character of the point T cannot occur without some other type of singular point between them, just as it is not possible for two mountains to rise from a landscape without some kind of valley between them. Now the series of equilibria met with in embryonic determination might be represented by a series of horizontal ledges descending from a point analogous to O, separated perpendicularly from each other by ridges or cols. Such a model, purely qualitative and intuitional, is represented in Fig. 11. The state of harmonious equipotentiality would then correspond to the summit point, where the instability is maximal, and a point could descend to the successive levels of instability not in one direction only, but in many, according to its position and other relations to the organiser region. This is to say, its prospective potency is not limited to its prospective significance. The mosaic egg would never occupy any of the higher ledges.[20]

20. Restriction of potentiality by multiple bifurcation finds an analogy in human technology in the automatic marshalling-yards

Inorganic solutions or vapours may exist in the condition of supersaturation, and they may then be brought to crystallise or to condense by the introduction of a suitable nucleus, the dimensions of which may be minute relative to the system as a whole. This state of affairs is called a metastable state. Thermodynamically it is defined by the fact that its potential is a minimum, but not an absolute minimum. It is not unlikely that the term metastable could justifiably be applied to the plateau-like states which occur in the above model in embryonic determination. The appropriate organiser would then correspond to the nucleus which ends the state of supersaturation in an inorganic system. It is reminiscent of Bertalanffy's phrase,[21] that the egg is "charged with morphological form" or has a "morphological charge." Just as inorganic metastable systems are stable in the absence of their nucleus, so, in the absence of the organiser,

of railways, as, e.g., at Whitemoor, near March, Cambridgeshire: cf. *Railway Gazette,* Sept. 1929. (*See* Fig. 45).

The equilibria here envisaged, which become ever more stable as development proceeds, should not be confused with other, also no doubt legitimate, uses of the concept in biology. Thus Przibram (*Anorganischen Grenzgebiete d. Biol.,* p. 64 ff. and *Journ. Exp. Zool., 5,* 1907, 259) speaks of the completed organic form of the animal as being an equilibrium which is disturbed by e.g. the removal of a limb. There ensues a struggle between growth and surface tension, ending in an equilibrium which *is* the normal form once more. This is put crudely, but the idea is rather crude.

21. Bertalanffy, L. v., *Kritische Theorie d. Formbildung* (Berlin, Bornträger, 1928), p. 223. "So, wie ein Akkumulator mit Elektrizität geladen ist, so ist gewissermassen das entwicklungsfähige Ei mit 'Form' geladen, die es in Wirklichkeit umzusetzen trachet." *See also* his "Tatsachen & Theorien d. Formbildung als Weg zum Lebensproblem," *Erkenntnis, 1* (1930), 361.

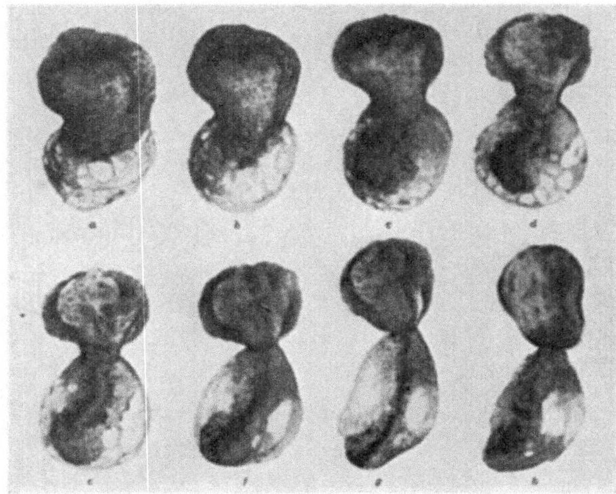

FIG. 12.

Successive stages in an exogastrulating newt embryo. The process here shown took 2½ days. The evagination instead of invagination of the endo-mesodermal mass resulted in complete absence of axial structures in the ectoderm. (From Holtfreter.)

FIG. 13.

Woodger's cone, illustrating embryonic development in a three-dimensional model. The long axis of the cone is time; the apex or origin represents the fertilised egg, and the area of cross-section at any given moment, the size of the embryo at that time. The origin of one group of cells from a single mitosis is represented by the black segment.

normal development and differentiation will not occur, as is beautifully seen in the exo-gastrulation experiments of Holtfreter[22] (Fig. 12).

To complete our picture of embryonic development on the basis of Fig. 11 we must remember that the obscure processes of invisible determination are going on side by side with the cleavage of the original cell into hundreds of cells, and with the increase of protoplasmic mass in the system at the expense of the inert yolk-materials, vitellin, phosphatides, etc., which were accumulated at the beginning. These processes have been represented by Woodger[23] under the form of a cone in which the long axis represents the dimension of time (Fig. 13). A plane of the cone perpendicular to its vertical axis represents in two dimensions the three-dimensional spatial organism conceived in abstraction from time. The apex of the cone represents the zygote. The black wedge in the model represents the system of cell-descendants of the cell k resulting in the establishment of the cellular component (organ) P. Woodger has shown how valuable this representation may be in the discussion of such questions as the relations between embryology and genetics.

22. Holtfreter, J., "Die totale Exogastrulation, eine Selbstablösung des Ektoderms von Entomesoderm," *Archiv. f. Entwicklungsmech., 129* (1933), 670; *also* "Organisierungstufen nach regionaler Kombination von Entomesoderm mit Ektoderm," *Biol. Centrlbl., 53* (1933), 404.

23. Woodger, J. H., "The Concept of Organism and the Relation between Embryology and Genetics, Pt. III." *Quart. Rev. Biol., 6* (1931), 178, pp. 193, 199.

Lotka, in his penetrating discussion[24] of the applicability of the principle of Le Châtelier to biological equilibria, concludes that even when the organism reacts in the direction of a restored equilibrium when exposed to a disturbing influence, the principle must be used with great caution. But the organism does not always act in this way. The ancient medical concept of *circulus vitiosus* exemplifies the point. A departure from equilibrium, instead of stimulating a compensating response, provokes a further departure in the same direction. The example is given of a person exposed to the hardships of adverse economic conditions; malnutrition lowers his resistance to bacterial infection; he contracts tuberculosis; there is loss of appetite, and the state of malnutrition may be accentuated and possibly fixed even if better conditions are provided. This amounts to a continuous accumulation of deviations from a stable equilibrium position, and in evolution-theory is known as the "cumulative cycle." It clearly cannot be a picture of embryonic determination, for that commences with unstable, not with stable, equilibrium; it is, in fact, the description of the reverse process. It raises the question whether this determination can be reversible.

There is some likelihood that this may be so. Driesch himself drew attention[25] to the case of the ascidian *Clavellina*, a form on which he made many investigations. It consists of two chief parts, the branchial apparatus and the intestinal sac; if these

24. Lotka, *loc. cit.*, p. 280 ff.
25. Driesch, Gifford Lectures, p. 94.

two parts are separated, each may regenerate the other in the typical way, by budding processes from the wound. But the branchial part may also lose the whole of its organisation and become a small white sphere, consisting only of epithelia corresponding to

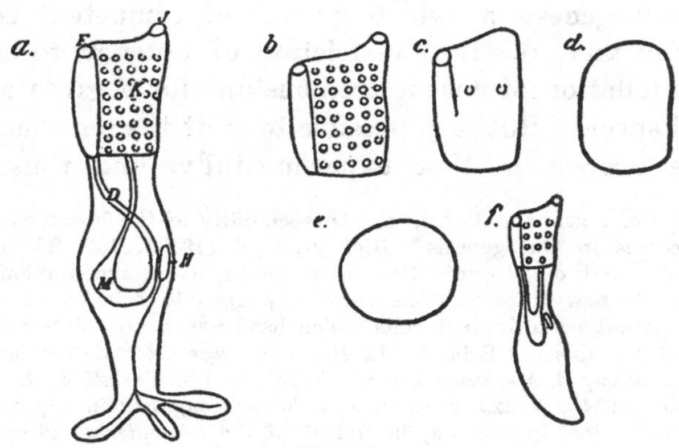

FIG. 14.

Driesch's experiments on the ascidian, *Clavellina*.
- a. Diagram of the normal animal: E and J, openings; K, branchial apparatus; D, intestine; M, stomach; H, heart.
- b. The isolated branchial apparatus.
- c, d, e. Different stages of reduction of the branchial apparatus.
- f. The new whole small ascidian.

the germ-layers, and then after a time a new organisation will appear. This new organisation is that of a complete, though small, ascidian (Fig. 14). Not only will the whole branchial part behave in this way, but also before its dedifferentiation it may be cut in any direction. This would seem to be a perfect case of de-

determination, but we know many instances of dedifferentiation and degrowth.[26] The subject has recently been discussed by Harrison.[27]

The determinative factor in differentiation can, it seems, be disengaged from the growth process just as much as the other factors can. The particular points in ontogenesis at which groups of competent cells suffer their destined restriction of potency form a constellation of characters constant for a given animal species. But it is possible to shift the determinative process in time experimentally, and thus to

26. *See* Needham, J., "On the Dissociability of the Fundamental Processes in Ontogenesis," *Biol. Rev., 8* (1933), 180. There is doubt about de-determination in ascidians, as in the opinion of some, the new differentiation which may ensue is to be ascribed to pluripotent or totipotent cells which have never been determined at all. On this *see* Schaxel, J., *Die Leistungen der Zellen bei d. Entwicklung d. Metazoen* (Berlin, 1915), and in *Verhdl. d. d. zool. gesell.* (1914), p. 122. Schaxel and Driesch engaged in a polemic on the subject in vols. 35, 36 and 37 of the *Biologischen Zentralblatt,* which may be recommended to those who wish to pursue the matter further. More recently, Spek (*Archiv. f. Entwicklungsmech., 111* [1927], 119) has described amoeboid cells of definite histological properties present in the adult ascidian, which collect in the winter-buds at the stolon and which are almost certainly responsible for the redifferentiation after the resorption process. Here there is organiser-action of a sort. Similar arguments apply to the undetermined character of regenerating amphibian limbbuds (*see* Weiss, P., *Morphodynamik* [Berlin, Bornträger, 1926], p. 16 ff.). Fischer, A., and Mayer, E. (*Naturwiss., 19* [1931], 849) also raise the general question whether de-determination does not accompany the dedifferentiation seen in explanted cells of adult organisms. Apparently no one has yet subjected explanted cells of adult amphibia to the action of organiser or evocator.

27. Harrison, R. G., "Some Difficulties of the Determination Problem," *Amer. Nat., 67* (1933), 306. Harrison is inclined to question the value of the determination-concept on the ground that determination may be reversible. But is the methodological value of the concept any the less on that account?

throw it out of gear with the growth process. Twitty[28] found that the polarity of the ciliary cells of the amphibian embryo is determined during the closure of the neural folds, for the cilia of ectodermal

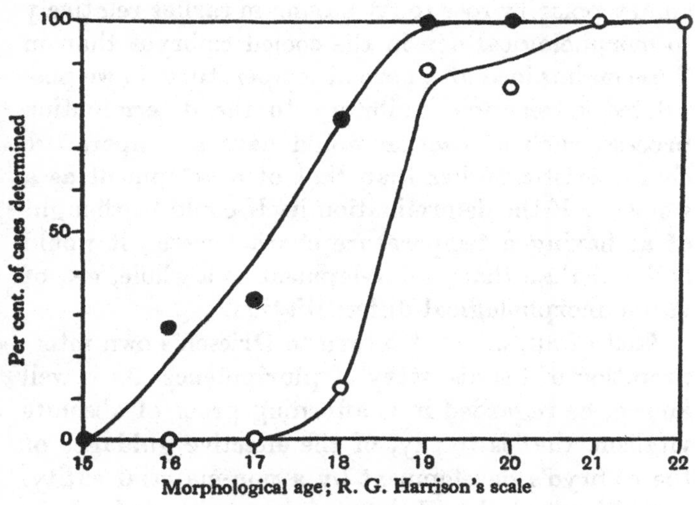

FIG. 15.

Determination of ciliary polarity in *Amblystoma* (experiments of Twitty). White circles represent embryos kept at 15°C. before the operation, black circles represent those kept at 7°.

grafts rotated through 180° before that stage beat in the same direction as the cilia of the adjacent host ectoderm, while ciliary grafts transplanted and rotated later retained their original direction of wav-

28. Twitty, V. C., "Experimental Studies on the Ciliary Action of Amphibian Embryos," *Journ. Exp. Zool., 50* (1928), 319.

ing. But in embryos allowed to develop at lower temperatures Twitty found that the ciliary polarity appeared at a much earlier stage, relatively to the morphological development. As Fig. 15 shows, the percentage of transplantations showing determined ciliary polarity rose to its maximum earlier relatively to morphological age in the cooled embryos than in those maintained at a normal temperature. If we postulated a reaction inhibitory to the determination process, such a reaction would have a temperature characteristic higher than that of development as a whole, or if the determination itself could be thought of as having a temperature characteristic, it would be lower than that of development as a whole, i.e., of visible morphological differentiation.

Last of all, we must return to Driesch's own interpretation of his discovery of pluripotence. As is well known, he regarded it as affording proof of absolute vitalism, that is to say, of the effective guidance of the embryo's development by a non-material entity. It will be remembered that reference was made above to the invariable factor of wholeness, E, which is involved in the formation of two whole embryos from two blastomeres which, left to themselves, would only have formed one. "It was not without design," wrote Driesch, "that I chose the letter E to represent this. Let that factor in life phenomena which we have shown to be a factor of true autonomy be called 'entelechy,' though without identifying our doctrine with what Aristotle meant by the word ἐντελέχεια. There is, however, at work a something which bears the end in itself ὃ ἔχει ἐν ἑαυτῷ τὸ τέλος."[29] The (Drieschian) entelechy was then defined as intensive manifoldness (*intensive Mannigfaltigkeit*) as opposed to the extensive or visible manifoldness which becomes evident as development proceeds.[30] All order in morphogenesis is exclusively due to the action of entelechy.[31] The entelechy does not exist in space (space-time), but acts into space,[32] it is therefore not localisable at any point in space-time.[33] Lacking all the characteristics of quantity, and being incapable of measurement, it cannot be a form of energy,[34] and it cannot therefore affect the first law of thermodynamics.[35] Its definition was finally admitted by Driesch himself to be "a mere system of negations"[36] analogous to the "negative way" of mystical theology. He could, however, describe its action as that of suspending possible reactions,[37] one or more of which would otherwise proceed in its absence, and in this sense it could be called an "arranging agent."[38]

It seems really astonishing to us to-day that Driesch should have felt himself impelled to have recourse so soon in his analysis to a conception so

29. Driesch, H., Gifford Lectures, p. 106.
30. *Idem,* p. 245. 31. *Idem,* p. 154.
32. *Idem,* pp. 254, 298. 33. *Idem,* p. 299.
34. *Idem,* p. 256. 35. *Idem,* p. 257.
36. *Idem,* p. 300. 37. *Idem,* p. 261.
38. Gifford Lectures, p. 292. In this connection it is important that the teleological significance of the entelechy was greatly diminished in the following decade by the discovery of self-differentiation and the realisation that all embryos pass through a stage where no regulation is possible. The entelechy thus deserts its post when the work is one third completed. Embryologists were accordingly led to question whether it had ever been there at all.

obscure in all its particulars, so fantastically out of harmony with the natural products of the scientific method, so certain to close the door to further experimentation, so exactly resembling the Galenic "faculty." We cannot see why it was necessary to place the intensive manifoldness of the egg outside the physical world, in view of the tremendous complexity which the colloidal constitution of protoplasm must involve. Even above the atomic level there would be ample room for it. It must have been evident enough, even in the last century, before the work of Köhler[39] and his school, that, when a magnet is divided into two, we are not left with one north pole and one south one, but that an immediate "regeneration" of the whole pattern takes place, and that we have two fields of force, similar morphologically to the original one, but smaller. The particles of the magnet are not, as it were, "determined" to form part of any given pole, but can form part of either, according to their position in the whole. And this analogy only needs transposition to the more complicated realm of biological colloids.

Again, the *Verschmelzungsversuche*, where two eggs are fused together, yet one normal embryo results, have their analogy in the long-known phenomenon of convergent crystallisation (*Sammelkristallization*). Spatially separated small crystals commonly

39. Köhler, W., *Die physische Gestalten*, 1919; *Jahresber. u. d. ges. Physiol.*, 1922; and "Zum Problem der Regulation," *Archiv. f. Entwicklungsmech.*, *112* (1927), 315. Already in 1877 Pflüger (*Archiv. f. d. ges. Physiol.*, *15* [1877], 61) had seen the significance of the magnet analogy for life phenomena.

unite themselves into large ones of the same structure. Still more striking is the behaviour of the ammonium oleate crystals of O. Lehmann.[40]

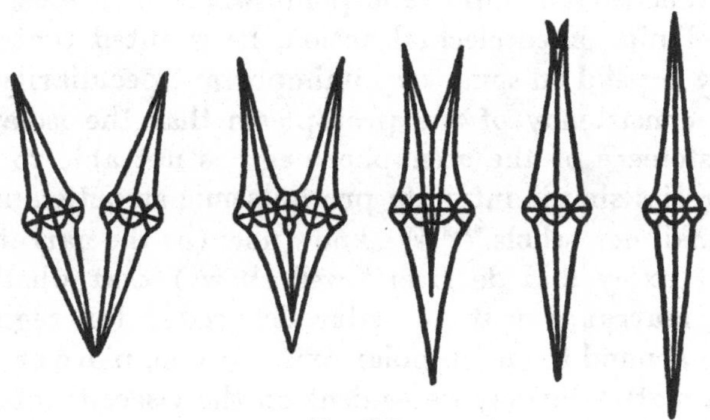

FIG. 16.

"Duplicitas" and fusion of form in crystals of ammonium oleate. (From Lehmann, after Rinne.)

These soft forms, if side by side, join up together, passing through all the stages corresponding to *Duplicitas* greater or lesser in animals.

Not seldom Driesch passed into pure dogmatism. He did not wish, he said, to discredit a thorough and detailed study of osmosis or colloid chemistry in their relation to morphogenesis. "But these investigations never give us a solution of the problem."[41] Or again "Atoms and molecules by themselves can

40. Lehmann, O., "Die Lehre v. d. flüssigen Kristalle u. ihre Beziehungen z. d. Problemen d. Biol.," *Ergebn. d. Physiol., 16* (1918), 256. H. Przibram gives in his book *Anorganischen Grenzgebiete d. Biologie* (p. 74 ff.) many instances approaching harmonic equipotentiality in crystals and paracrystals.

41. Driesch, H., Gifford Lectures, p. 63.

only account for form arranged, so to speak, according to spatial geometry, they can never account for form such as the skeleton of the hand or foot."[42] But he touched an important point when, in discussing the limits of entelechial action, he granted that "it may depend on some very unimportant peculiarity in the consistency of the protoplasm that the isolated blastomere of the ctenophore egg is not able to restore its simple intimate protoplasmic structure into a small new whole."[43] We know now (as the narrative of Huxley and de Beer[44] well shows) that whether the rearrangement of materials needed for regulation around the main polar axis can take place or not is a matter largely dependent on the viscosity of the protoplasm—an eminently physical factor. The distinction between mosaic and regulation eggs, therefore, can partly be defined in terms of internal friction.

The intensive manifoldness above the atomic level in the uncleaved egg generally receives to-day a formulation in terms of the gradient concept, Child's great contribution to biology. The polarities of the egg, brought into being partly by its chance position in the ovary, partly by the point of entry of the spermatozoon, and other similar factors, involve a system of gradients describable in a three-dimensional net of co-ordinates. Quantitative differences in activity along the gradients lead to qualitative differentiation; localisation is due to relative position on a

42. *Idem*, p. 101. 43. *Idem*, p. 266.
44. Huxley, J. S., and de Beer, G. R., *loc. cit.*, p. 126, etc.

gradient. Even the best cases of harmonious equipotential systems possess limitations unrealised by Driesch; thus the single blastomere isolated from the first four of the sea-urchin regulates because it possesses the full extent of the gradient, for if it is divided transversely, or if the original whole egg is divided transversely (equatorially), no perfect embryo will be formed.[45] Perfect equipotentiality therefore depends on the possibility of separating a system into parts without segregating regions of different potencies.[46]

Driesch's proposition "that a system, in the course of becoming, is unable to increase its manifoldness of

45. Hörstadius, S., "Studien über die Determination bei *Paracentrotus lividus*," *Archiv. f. Entwicklungsmech., 112* (1927), 239. "Uber die Determination des Keimes bei Echinodermen," *Acta Zoologica, 9* (1928), 1; Plough, H. H., "Defective Pluteus Larvæ from Isolated Blastomeres of *Arbacia* and *Echinarachnius*," *Biol. Bull., 52* (1927); Schaxel, J., *Die Leistungen der Zellen bei der Entwicklung der Metazoen* (Jena, 1915); Conklin, E. G., *Amer. Nat., 67* (1933), 289.

46. It is of great interest that Driesch's argument, though leading to such unmechanistic conclusions, is fallacious precisely because it adopts such mechanistic premises. To these it *adds* entelechy. The fact that a heap of cells (*summenhafte Gesamtheit*) becomes a whole, in spite of any desired removals and injuries, can only be explained by the entelechy hypothesis, says Driesch. But the premise is not correct; the regulating egg is not a heap of cells (*summenhafte Beieinander*), but an organised system. Regulation can only occur if the system is in a position to reconstitute its original pattern after the removal or shuffling of its cells. The unorganised heap of cells is an abstraction which does not enter into the question, and can be banished from discussion. But if it goes the entelechy goes with it. (Bertalanffy, article in *Erkenntnis, loc. cit.*, p. 368.) In 1927 Driesch is still writing: "Der Organismus kann zwar in jedem Moment seines Daseins seiner materiellen Seite nach als chemisch-aggregative 'Maschine' gekennzeichnet werden, aber seine andere Seite, der Maschinenbenutzer, darf nicht

itself"[47] raises, of course, the whole of the age-long controversy between the preformationists and the epigenesists. But Woodger[48] points out that an organised entity having components standing in internal organising relations to each other *must* have its degree of multiplicity increased if (1) the number of components is increased, (2) the complexity of the relations in which those components stand to one another is increased, (3) the intrinsic patterns of the components become different from one another. Thus, granted the possibility of spatial repetition of a pattern and the possibility of "differentiating division" and "histological elaboration," there is no need to appeal to agents of any kind. If we do not need to invoke the entelechy to account for cell-division, this alone will suffice in principle, since this process must yield an increase in multiplicity in succeeding slices

vergessen werden." (Driesch, H., "Uber neuere Vitalismuskritik," *Biol. Centrlbl.*, *47* [1927], 651). This is the purest Cartesianism; it was Descartes who introduced the practice of calling organisms machines, and then postulating transcendent mechanics to drive them.

47. Driesch, H., Gifford Lectures, p. 319. This is clearly put forward in his *Der Begriff der organischen Form* (Berlin, Bornträger, 1919), p. 42 ff., where we find: "Grundsatz I: Wie im rein Logischen die Folge nicht mannigfaltiger an (logischem) Inhalt sein kann als der Grund, so kann sich auch im Laufe des Werdens der Grad der Mannigfaltigkeit eines Natursystems nicht 'von selbst' erhöhen. Wer das zulassen würde, würde den Begriff der Kausalität als einer Analogie zur Konsequenz, verletzen." But is there any justification for this unexplained extension of formal logic to natural phenomena? If it is wrong, Driesch's whole theoretical system is wrong.

48. Woodger, J. H., "The Concept of Organism and the Relation between Embryology and Genetics, Pt. III," *Quart. Rev. Biol.*, *6* (1931), 178, p. 202. *See also Biological Principles*, chap. ix.

of the life-history (*see* Fig. 13). "Neither the preformationists nor the epigenesists knew anything of spatial repetition of pattern or of histological elaboration, since such processes occur only in living cells, and there was therefore nothing in the experience of the early embryologists to suggest that such things occurred." Here Woodger goes to the bottom of the embryological problem in its biochemical aspect, for our aim must be to discover how it is possible for the formation of duplicate protein (and other) molecules to occur. The mass-production of replicas is the issue. Whether it occurs by some form of template method or, as seems more probable, by the scission into units of an endless protein chain, we have at present no idea.

In his acute analysis of causation, Woodger[49] has another argument which has an important bearing on the theory of the harmonious equipotential system. If two entities are manifestly, i.e., observably, non-different, we cannot infer that they are not intrinsically non-different. This is what makes causal analysis so difficult, because it means that in any biological experiment there will always be an element of doubt in the assumption that the two biological objects compared or their environments are intrinsically non-different. That two entities are *not* intrinsically different is *always* an assumption. Thus in the case of equipotence, it is argued that, since any given

49. Woodger, J. H., "Some Apparently Unavoidable Characteristics of Natural Scientific Theory," *Proc. Aristot. Soc., 32* (1932), 95, p. 107.

part of the original egg can become any given part of the resulting embryo, therefore the spatial parts of such eggs are intrinsically non-different. In that case it is impossible that differences can subsequently arise in development, since the environment cannot be alone responsible. Hence we are driven to conclude that an entity, the entelechy, which is not a spatial part of the egg, is "at work" producing the differences required for further development. Now it seems clear that such arguments can never be decisive on account of the principle that intrinsic non-difference is always assumed, and consequently it is always open to an unbeliever to say that such experiments do not exclude the possibility of intrinsic differences between the parts concerned *which are not revealed when they are in isolation.* Or, in more biological terminology, the properties of a given part other than its fate may well be different according to whether it forms part of a two- or more) cell system, or of one of the two cells in isolation.

The meaning of the phenomena of equipotentiality has recently been the subject of an interesting polemic between Mirakel[50] and Wermel.[51] Mirakel purported to bring a proof of the existence of the entelechy logically better than Driesch's, but Wermel had little difficulty in showing that his argument was

50. Mirakel, R. C. (a pseudonym?), "Zur Beweisung d. Vitalismus, offener Brief an Hans Driesch," *Biol. Centrlbl., 53* (1933), 614.

51. Wermel, J., "Über die Beweisung d. Vitalismus, Antwort an R. C. Mirakel," *Biol. Centrlbl., 54* (1934), 313.

vitiated by a failure to take account of the dominant position which one part of the egg or egg-fragment (the organiser region) has over the remainder. Mirakel considered that better support for the Drieschian hypothesis could be derived from the effects of "*Verlagerunsversuche*,"[52] i.e., experiments in which the primary blastomeres of the egg are shuffled at random together and yet give a normal embryo, rather than from the effects of fusion, amputation, or isolation. Some non-spatial agency must be acting upon the shuffled blastomeres in a marshalling manner. But Wermel showed that this conclusion was only true if the material factors which determined the fates of the parts had to act on each one independently and separately; it was not true if one part could determine the fates of the rest. This would then happen no matter what chance arrangement of the blastomeres had taken place. And indeed we know that without the presence of an organiser in the isolated fragment, there can be no harmonious equipotentiality.[53]

In conclusion, a word may be said of the remark-

52. Driesch, H., *Archiv. f. Entwicklungsmech., loc. cit., 4* and *14*.

53. The whole of this argument was foreseen in a somewhat abstract way by Woodger ("The Concept of Organism," Pt. I, p. 20). This is a summary of his reasoning: The possibilities before us in the matter of embryonic development are:

1) a Weismannian nuclear preformation,
2) a cytoplasmic preformation,
3) environmental differences during development,
4) appeal to a Drieschian transcendent principle,
5) rejection of the causal postulate.

Of these 1 and 3 are excluded as decisively as anything can ever be by experiment. Even Driesch shrinks from 5. No real help is

able animistic and anthropomorphic tendency noticeable in Driesch's concept of an "arranging agent." For centuries science has struggled to rid itself of the remains of popular demonology.[54] When Cambridge got its first Professor of Chemistry[55] (in 1703) acids were male and alkalies female, minerals grew, like plants, from seeds, slaked lime protested by giving out heat, and solid bodies cleaved to themselves in cohesion because they preferred the touch of the tangible more than the feeble contact of air. The entelechy shamelessly belongs to pre-Boylian biology. The work of the logistic positivists, Carnap, Schlick, and Neurath, brilliantly illustrates this.[56] Their purification of scientific terminology convinced them that the majority of biological concepts, such as cell-division, growth, regeneration, etc., could be reduced to incorporation in the "physical language," i.e., in language based on direct experience. This was not

forthcoming from 4. Only cytoplasmic preformation remains. "Would it be possible to appeal to just enough cytoplasmic preformation to give us our main axes and the primary germ-layers, and could we thenceforwards appeal on a large scale to relational properties?" (in which organiser phenomena would be included). This is certainly the outcome, in general terms, of modern embryology.

54. *See* the interesting article of Gregory, J. G., "The Animate and Mechanical Models of Reality," *Journ. Philos. Stud., 2* (1927), 301.

55. John Francis Vigani, *see* Peck, E. S., *Proc. Camb. Antiquar. Soc., 34* (1934), 34.

56. *See* Carnap, R., *The Unity of Science, tr.* M. Black, Psyche Miniature Series (London, Kegan Paul, 1934), p. 68 ff., and the trenchant essay of Winterstein, H., "Kausalität u. Vitalismus vom Standpunkt d. Denkökonomie" (Berlin, Springer, 1928).

the case, however, with words like 'entelechy,' which are incapable of formally correct definition, and hence can only occur in meaningless statements.[57] Schlick's prediction[58] that in the future no more books will be written about philosophy but that all books will be written in a philosophical manner, implies that the concept of entelechy will be searched for in them in vain. And the judgment of another philosopher on the entelechy is worth quoting. "There is no magic in mind as such," says Broad, "which will explain teleology; a mind does not account for anything until it has wit enough to have designs and will enough to carry them out. If you want a mind that will construct its own organism you may as well postulate God at once, for, if he cannot perform such a feat, it is hardly likely that what has been hidden from the wise and prudent will be revealed to entelechies."[59]

We must now return to study the course which our knowledge of embryonic determination took, apart from the entelechial speculations of Driesch himself.

When the mind is exhausted by too prolonged a contemplation of morphological form itself, either at the upper or lower levels, it may sometimes be a relief to turn to another method of penetrating into its

57. Carnap, R., "Überwindung d. Metaphysik durch logische Analyse d. Sprache," *Erkenntnis, 2* (1931), 219.

58. Schlick, M., *Communications to the VIIth International Congress of Philosophy* (Oxford, 1930).

59. Broad, C. D., "Mechanical Explanation and its Alternatives," *Proc. Aristot. Soc., 19* (1919), 86, p. 123.

nature, that is to say, the study of the influences which act upon it. Of the first importance among such agencies are the true morphogenetic hormones.[60]

Huxley points out that the classical concept of hormone can now be divided into several logical subdivisions.[61] First of all there are the metabolites with physiological functions, transported by the circulating blood, e.g., CO_2 or lactic acid. These, without possessing any high degree of specificity, certainly co-ordinate the activities of the body. Secondly, there are the true hormones,[62] organic molecules specially manufactured in *ad hoc* glands, and possessing great specificity at their sites of action, which again they reach transported by the blood. Thirdly, there are the "diffusing hormones," such as the substances described by Parker,[63] which appear to be liberated at the end-organs of nerves and to control the opening and closing of chromatophores. In this class would

60. Under the term "dependent differentiation" (the stimulating action of one organ upon another) the concept originates with Wilhelm Roux. It was much elaborated by Curt Herbst in a classical series of papers "Über die Bedeutung d. Reizphysiologie für d. kausale Auffassung von Vorgängen in der tierischen Ontogenese," *Biol. Centrlbl., 14* (1894), 657, 689, 727, 753, 800, *also 15* (1895), 721, 753, 792, 818, 849. Here the place of "formative stimuli" in development was discussed, largely from analogy with such processes as gall-formation.

61. Huxley, J. S., *Nature* (1934), and in a letter to the author.

62. According to local tradition, the word "hormone" was born in the hall of Caius College. Schäfer or Starling was brought in to dine by Hardy, and the question of nomenclature was raised. W. T. Vesey, an authority on Pindar, suggested ὁρμάω and the thing was done.

63. Parker, G. H., *Humoral Agents in Nervous Activity* (Cambridge, University Press, 1932).

come the substances of the nature of acetylcholine, shown by Dale and his colleagues[64] to be generated at nerve-endings. Fourthly, come the "contact hormones," i.e., the numerous organisers of different grades which determine the fates of the various parts of the embryo. The distinction between these last two classes is not very clear-cut, for the necessity of contact between the active and the passive tissues may only mean that the substance involved possesses very feeble powers of diffusion. This would be expected if it were of a fatty or lipoidal character.

A theoretical possibility is of interest here. It does not seem so far to have been envisaged that a substance might be a hormone without diffusing at all. It is possible to picture a single molecule or a molecular aggregate (perhaps of a paracrystalline nature) exerting an influence around itself in all directions of space for a considerable distance, even into microscopic dimensions if the argument of Hardy (*see* p. 134) is taken into account. The concentric zones which would then exist would not be zones of decreasing concentration of hormone (as we now generally assume) but zones of increasing randomness of arrangement of other particles. The gradient in a given direction would not be due to the falling off of the amount of the substance but to the falling off of its orienting arrangement. I shall refer again to this possibility in connection with the theory of the biological field.

64. e.g., Dale, H. H., and Feldberg, W., *Journ. Physiol.*, *82* (1934), 121.

We have been speaking of organisers as morphogenetic hormones, but we have not so far given the proof that this term can correctly be applied to them. In order to do this, it is necessary to make a short historical picture of the development of our knowledge of the amphibian primary organiser, i.e., that concerned with the determination of competent ectoderm to form neural plate and axial structures.[65]

It was a turning-point of the highest importance in the history of embryology[66] when in the spring of 1924 H. Spemann and Hilde Mangold obtained the first induction by the organiser of an amphibian egg. The dorsal lip of the blastopore of an embryo of the newt *Triton cristatus* was transplanted into the indifferent ectoderm of an embryo of *Triton taeniatus* at the same stage of development, i.e., the early gastrula. There it did not behave as presumptive medullary plate or epidermis would have done, but on the

65. Detailed references for the statements in the succeeding paragraphs will be found in the paper of Needham, J., Waddington, C. H., and Needham, D. M., *Proc. Roy. Soc., 114* (1934 B), 393, and in the book of Huxley and de Beer, *loc. cit.*

66. The original discovery of the Organiser has been attributed by some to W. H. Lewis and by E. W. MacBride to himself (*Discovery, 15* [1934], 218). Lewis (*Amer. Journ. Anat., 7* [1907]) implanted the dorsal lip of the blastopore into the ear region of older larvæ (*Rana palustris*). Notochord, neural tube, and somites resulted, but there was no influence on the host tissues. MacBride (*Proc. Roy. Soc., 90* [1918 B], 323) observed that the hydrocoele in sea-urchin larvæ "induced" the formation of the echinus rudiment. He compared this beautiful case of dependent differentiation with that of the amphibian lens, but obscured the hormonic significance of the result by phylogenetic speculations. It has been said: "Man kann wohl eine Entdeckung machen ohne es zu wissen, aber kaum ohne es zu wünschen."

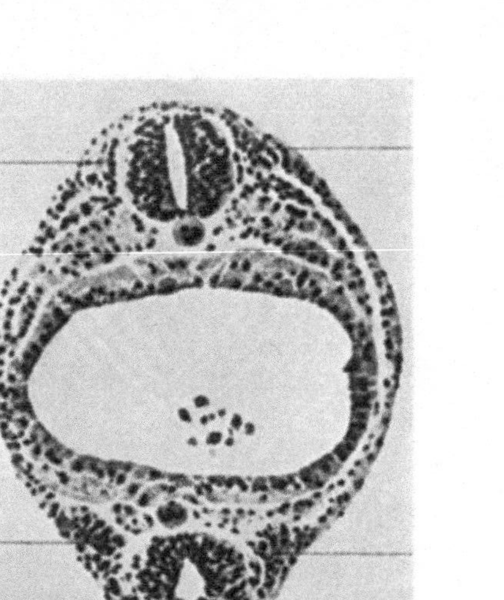

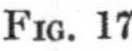

FIG. 17.

Example of a good induction by living organiser from the dorsal lip of the blastopore, primary embryo above, secondary embryo below. (From Spemann and Mangold.)

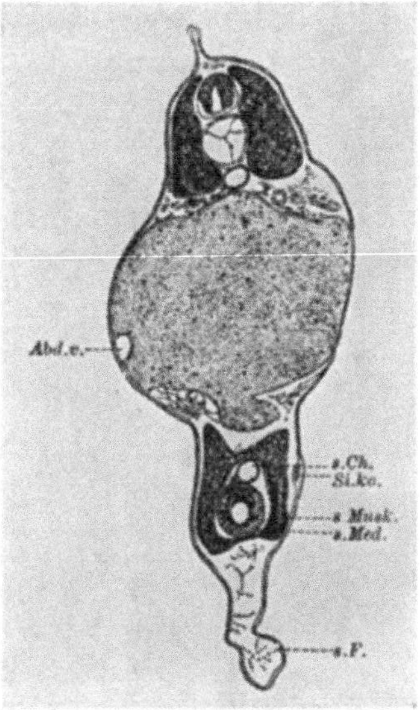

FIG. 18.

Induction by organiser narcotised completely by trichlorbutylalcohol. Primary embryo above, secondary embryo below. (From Marx.)

Abd.v. Abdominal vein.
s.Ch. Induced notochord.
Si.ko. Sensory bud.
s.Musk. Induced muscle.
s.Med. Induced neural tube.
s.F. Induced tail-fin.

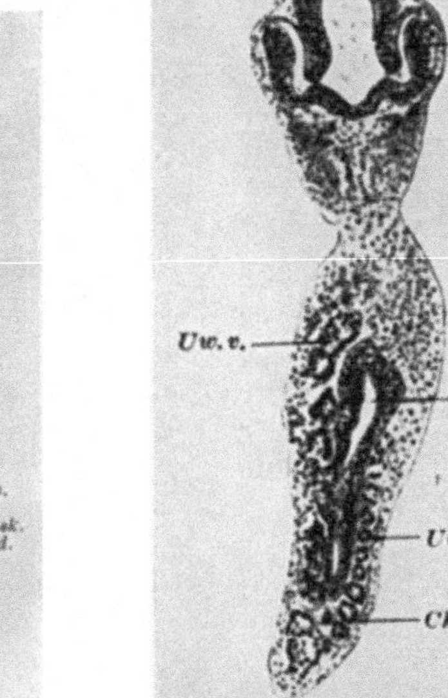

FIG. 19.

Induction by organiser in which the cellular structure has been destroyed by crushing. Primary embryo above, secondary embryo below. (From Krämer.)

Oc.pr. Host's optic cups.
Uw.v. Anterior somites.
Mr.sec. Induced neural tube.
Uw.sec. Induced somites.
Ch.sec. Induced notochord.

contrary asserted itself in its new environment and induced the formation of a new embryonic axis. Two embryos were thus produced on an egg which normally would have formed only one, for the induced structures included neural tube, notochord, auditory vesicles, mesoblastic somites, and Wolffian ducts. A typical result is shown in Fig. 17. Spemann, who had previously deduced the existence of such a centre in the dorsal lip of the blastopore, "radiating" its determining influence over the rest of the egg, called it the "organisation-centre."

For this fundamental discovery two main prior achievements had been necessary, firstly, the mapping of the normal fate of every part of the amphibian egg (Vogt) and, secondly, the technique of transplantation (Spemann). In this way the fruitful possibility of finding out what divergence from normal behaviour would take place when a part was placed in an entirely new situation was thrown open. As we have seen already (p. 55 ff.), there were found to be two main sorts of behaviour. Either the piece fell into step with its new environment, or it did not. Induction by an organiser could be regarded as the extreme case of the latter alternative, since the transplanted piece not only fulfilled its prospective significance, but also exerted a marshalling action upon cells within its sphere of influence. It was soon recognised that organiser phenomena were not confined to amphibia, for parallel cases were found in echinoderms (Hörstadius), insects (Seidel), and birds and mammals (Waddington).

The nature of the organiser influence was from the first recognised to set a problem the solution of which would profoundly affect our picture of the process of development. It meant nothing less than the discovery of the nature of the relational factor (*see* p. 77, note 53) in development. Various possibilities were discussed. It was tempting to think of the organiser region as the dominant end of a Childian gradient, the "high metabolic activity" of which would confer upon it its organising power. Unfortunately, this phrase has always had something of the flavour of a mystical symbol about it, and, although there is a moderate amount of evidence showing that gradients in the adult organism may be respiratory, almost none exists for the embryo. Moreover, since metabolism is almost synonymous with life, an explanation along these lines does not take us very far. Then there were the electrical effects, the potential differences, etc., which can be demonstrated on the amphibian egg. Their interpretation, however, is at all times so difficult that they could hardly be called upon to explain the phenomenon of the organiser. Still more unsatisfactory, if not positively quixotic, was the appeal to mitogenetic rays. The simplest explanation would be that induction was due to a single definite chemical substance acting in an almost endocrinological manner upon the competent ectoderm. This possibility invited a straightforward test.

The first important news was contained in the announcement of Spemann in 1931 that the organisation-centre would still induce after destruction of its

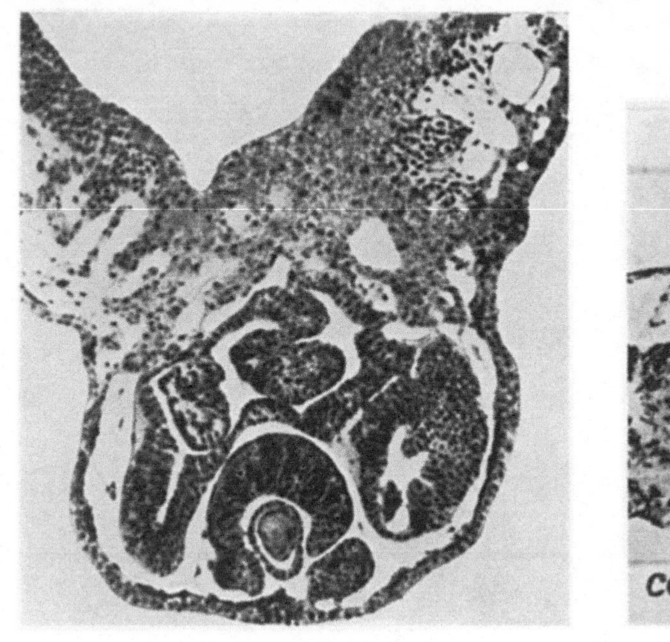
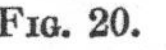

Fig. 20.

Induction by boiled organiser; only the secondary structures are shown. (From Holtfreter.)

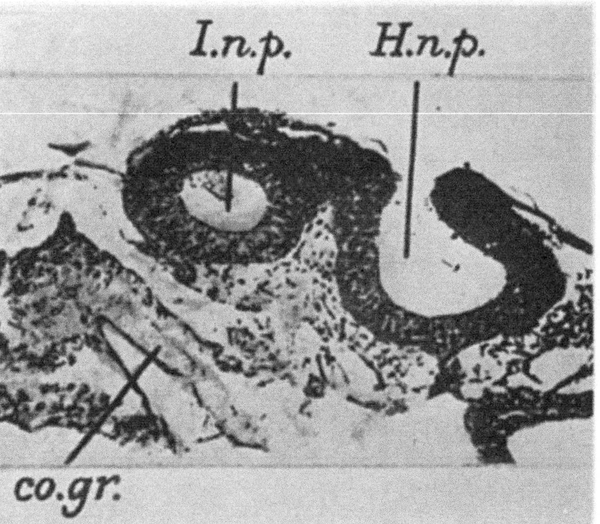

Fig. 21.

Induction by boiled organiser in the chick; a cross-section through the blastoderm. (From Waddington.)

co.gr. Boiled organiser graft.
I.n.p. Induced neural tube.
H.n.p. Host's neural tube.

cells. Pieces of tissue possessing the power of induction were crushed between glass plates, yet when implanted, such disorganised material could organise nearly all the constituent morphological parts of a secondary embryo (Fig. 18). In the same year Marx showed that the inducing power was fully retained when the cells of the dorsal lip were completely narcotised (Fig. 19). Such experiments strongly indicated that the organiser was chemical in nature.

In the following year, four workers simultaneously (Bautzmann, Spemann, Holtfreter, and Otto Mangold) showed that the organiser was remarkably resistant to heat, since it could be boiled for some time without loss of activity (Fig. 20). This definitely excluded the difficult possibility that the organiser was some delicate happening involving enzyme action. The stability of the organiser to heat was demonstrated also for the chick by Waddington (Fig. 21). At least as important, however, was the really remarkable discovery of Holtfreter, that parts of the embryo, which do not possess inductor capacity in the living condition, acquire it when they are boiled or when their proteins are denatured by immersion in alcohol, acetone, etc. Thus endoderm of the early gastrula induces strongly after being killed, but never before.

Next year (1933) the first experiments designed to test the activity of true cell-free extracts of the organiser region were made. In Berlin-Dahlem it was shown that the organiser activity could still be manifested after the crushed cell-debris had been spun out of the extract by centrifugation (Needham, Waddington, and Needham) (Fig. 22). The embryos were then ground up with anhydrous sodium sulphate, to absorb all water, and thoroughly extracted with solvents such as ether and petrol-ether. Again activity was found in the extract (Fig. 23), strongly indicating that the organiser hormone, for such it could now legitimately be called, was of a fatty, lipoidal, or sterolic character, and so soluble in organic solvents. During the spring of 1933 others were also at work. In Freiburg Spemann, Fischer, and Wehmeier simultaneously could show the presence of the organiser in cell-free extract. Holtfreter at Dahlem, in a long series of experiments, proved not only that the organiser hormone was present in most of the adult tissues of the newt (Fig. 24), but also that it could be found throughout the animal kingdom, in animals as far apart as *Daphnia* and man. This was the crowning demonstration of the complete lack of species-specificity of the organiser. In the light of Holtfreter's work the simultaneous inductions in the newt by chick primitive streak (Hatt at Paris) and by human cancer (Woerdemann at Amsterdam) were less striking than they would otherwise have been. It could only be concluded that the organiser was one or more chemical substances present throughout the animal kingdom. And the significant fact emerged that the only situations where it was necessary to kill the cells in order to unmask it were precisely those where in normal development the organiser would be harmful. Only in the non-induc-

CHEMICAL EXPERIMENTS ON EMBRYONIC INDUCTION (NEWT)

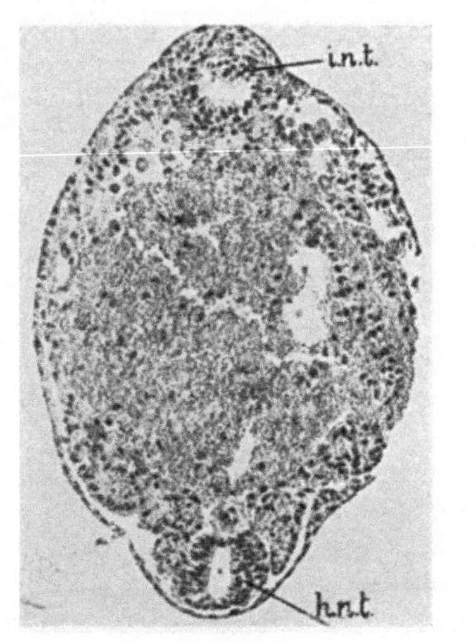

FIG. 22.

Induction by cell-free extract of neurulæ. (From Needham, Waddington, and Needham.)
h.n.t. Host neural tube.
i.n.t. Induced neural tube.

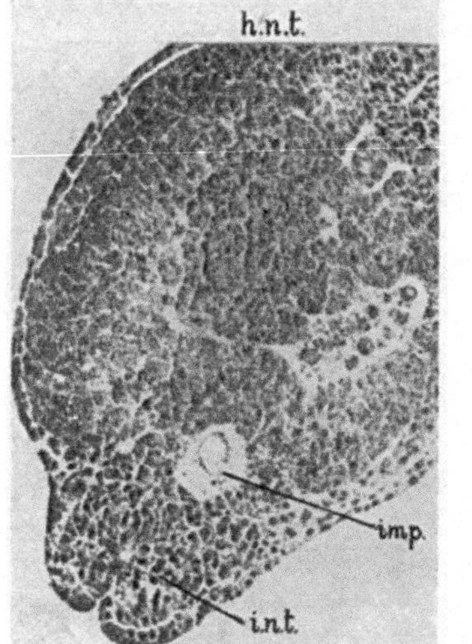

FIG. 23.

Induction by ether extract of neurulæ. (From Needham, Waddington, and Needham.)
h.n.t. Host neural tube.
i.n.t. Induced neural tube.
imp. Implant.

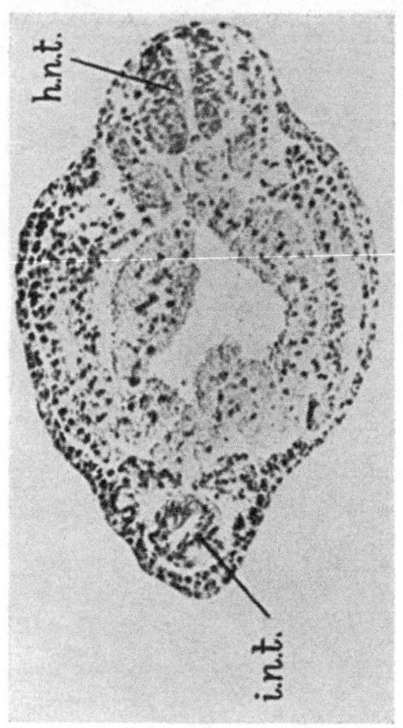

FIG. 24.

Induction by liver-tissue of adult newt (unboiled). (From Needham, Waddington, and Needham.)
h.n.t. Host neural tube.
i.n.t. Induced neural tube.

ing parts of the normal egg is the organiser masked or in some way inactivated.

In this fruitful year Fischer and Wehmeier also reported that they had obtained inductions in the newt by the implantation of glycogen. This fact could not be reconciled with the finding that the organiser was ether-soluble, if the glycogen was pure, but in 1934 it was shown that glycogen as ordinarily prepared contains as an impurity the organiser in highly active form (Waddington, Needham, Needham, Nowiński, and Lemberg) (Fig. 25). Now, too, the next step was taken in showing that the active substance was in the unsaponifiable fraction (Fig. 26).

At this point, the subject began to look as if it might link up with a branch of biochemistry in which greater progress has been made in the past few years than in any other. I refer to the biochemistry of the sterols.[67] The earlier researches of Wieland and Windaus provided a great mass of factual data about the chemical relationships of cholesterol and its related substances, but not until the brilliant suggestion of a new fundamental formula by Rosenheim and King did the subject become a logical and coherent whole. Then almost simultaneously it became clear that a sterol-like structure was common to several groups of compounds of extreme biological importance, (a) the sex-hormones both male and female (Dodds, Butenandt, Marrian, etc.), (b) the cancer-

67. *See especially* the review of Rosenheim, O., and King, H., *Ann. Rev. Biochem.*, *3* (1934), 87.

producing hydrocarbons, both naturally occurring in tar and synthetic (Kennaway, Cook, Hieger, etc.), (c) vitamin D, irradiated ergosterol (Bourdillon, Callow, etc.), (d) the aglucones of the pharmacologically active cardiac glucosides (Jacobs, Tschesche, etc.). At the present time, the greatest caution is wisely being exercised about the relationships of these facts, and there is a reluctance to speculate too freely along the inviting line, for instance, that cancer may be a metabolic disease affecting the metabolism of the sterols. Nevertheless, it is a remarkable coincidence, in view of these facts, that the amphibian organiser appears in the unsaponifiable fraction of the ether extract of an adult organ such as newt liver. To test its nature further, precipitations with digitonin have been made, with positive results (Waddington, Needham, Nowiński, and Lemberg) (Fig. 27). This does not prove that the organiser is a sterol, but the digitonin precipitation is not one likely to adsorb considerable amounts of a non-sterolic substance. Still more striking evidence was found in the work of Waddington and Needham, who implanted some of the synthetic hydrocarbons prepared by Cook, and obtained the induction of perfect neural tubes in the newt gastrula (Fig. 28).

These facts, and others to be mentioned in a moment, force us, as Dodds has pointed out,[68] to revise our conceptions of hormone specificity. Whereas in the classical conception, the essence of hormone ac-

68. In his masterly Goulstonian Lectures "Hormones and Their Chemical Relations," *Lancet* [1934 (i)], 931, 987, 1048.

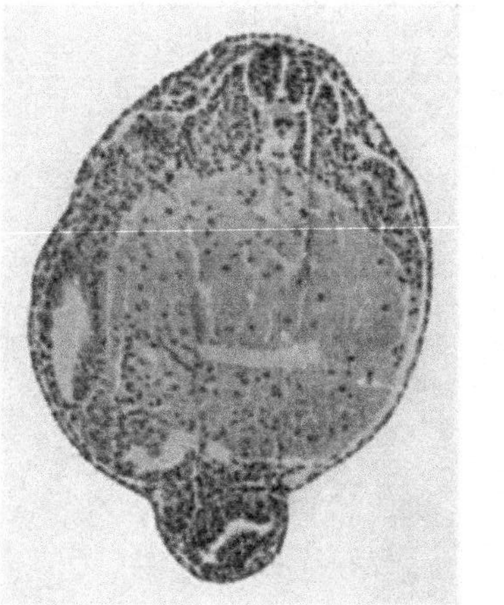
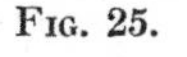

Fig. 25.

Induction by ether extract of crude glycogen prepared according to the classical method of Pflüger. (From Waddington, Needham, Nowiński, and Lemberg.)

h.n.t. Host neural tube.

i.n.t. Induced neural tube.

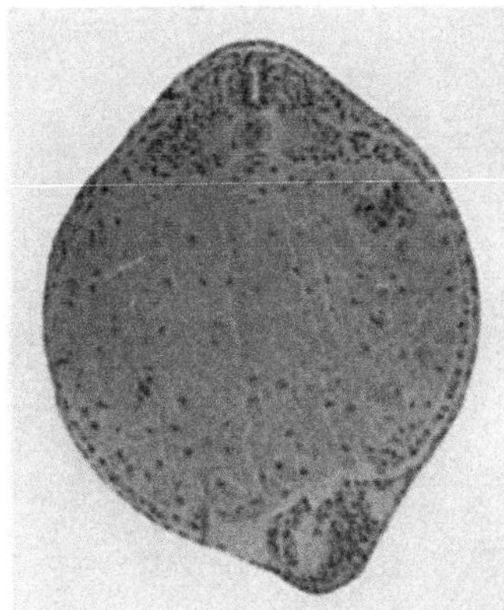

Fig. 26.

Induction by the digitonin-precipitable fraction of the unsaponifiable material of the ether extract of mammalian liver. (From Waddington, Needham, Nowiński, and Lemberg.)

h.n.t. Host neural tube.

i.n.t. Induced neural tube.

tion, such as that of adrenalin or thyroxin, was specificity, we know now of many cases where quite a large

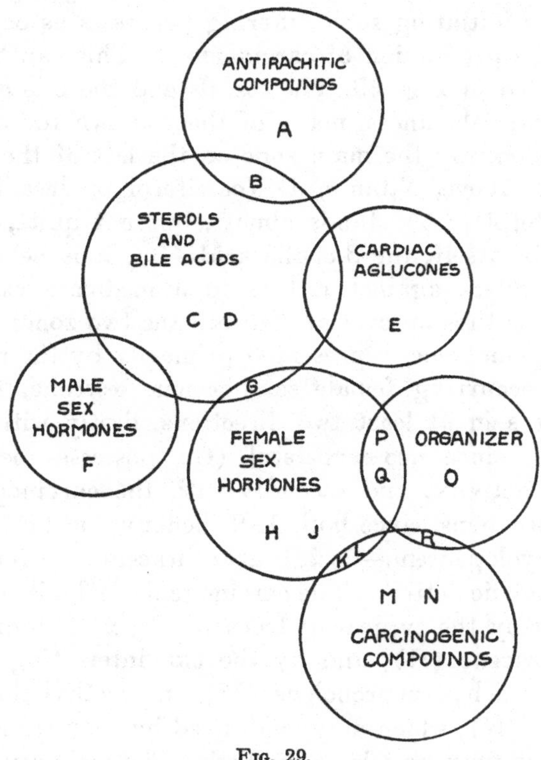

FIG. 29.

Chart to illustrate the relationships between the biologically important compounds of the phenanthrene series.

change in the molecular structure may be made, without materially altering the biological activity of the substance. "Not only," writes Dodds, "have we now already proved the lack of specificity of hormones, but we have proved that a single substance of known constitution may have at least two entirely separate actions initiating such differing processes as oestrus and the production of carcinoma."[69] This can be illustrated in Fig. 29. The sterols and the bile acids, classical substances, many of them known for many years, occupy the main zone on the left of the diagram. Above, Vitamin D (calciferol or irradiated ergosterol) (A) stands almost, but not quite, in a class by itself, for β-cholic acid (B), it is believed, will protect against rickets to a mediocre extent. There is thus an overlap between the two zones. The œstrogenic zone, represented primarily by the naturally occurring female sex-hormone œstrone, (H), overlaps in at least two directions, firstly with the sterols, since neo-ergosterol (G) possesses oestrogenic activity, and secondly with the carcinogenic hydrocarbons, since both 1, 2, benzpyrene (K) and 5, 6, cyclopenteno—1, 2, benzanthracene (L) possess oestrogenic activity. The carcinogenic realm is represented by the typical hydrocarbon 1, 2, 5, 6 dibenzanthracene (R), and by the two interesting substances dehydronorcholene (M) and methyl-cholanthrene (N), which may be derived by easy transformations from sterols in the body. The male sex-hormone and the pharmacologically important cardiac aglucones are closely related substances. Now into this scheme comes the organiser of such interest to embryologists. Whatever the naturally occurring

69. *Loc. cit.*, p. 990.

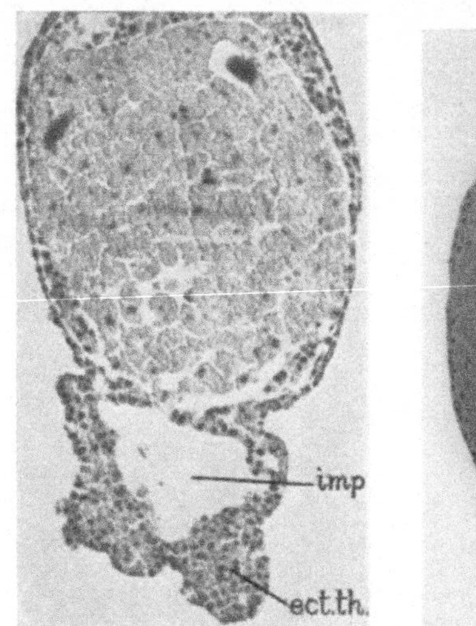

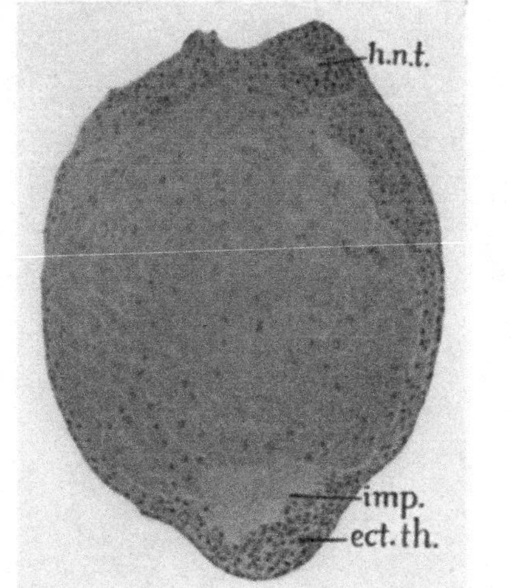

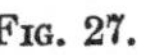

Fig. 27.

Absence of activity from the non-digitonin-precipitable fraction of the unsaponifiable material of mammalian liver (right). Absence of activity when a sterol such as pure cholesterol is implanted (left). In both cases there is a proliferation of cuboidal ectoderm cells, but no differentiation, histological or morphological. (From Waddington, Needham, Nowiński, and Lemberg, and from Needham, Waddington, and Needham.)

h.n.t. Host neural tube.
imp. Implant.
ect.th. Ectodermal thickening.

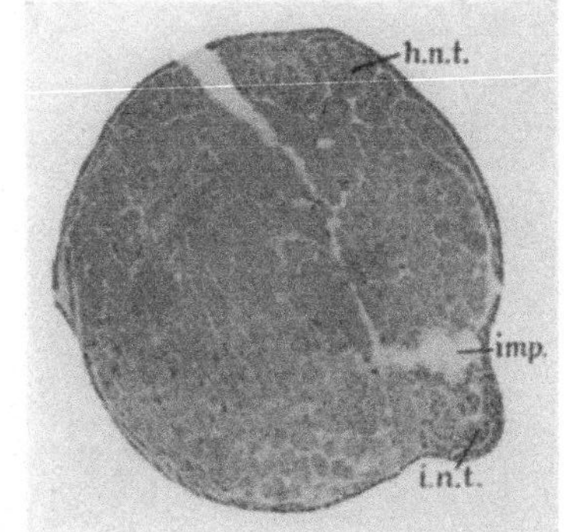

Fig. 28.

Induction by the synthetic hydrocarbon 1, 9, dimethyl-phenanthrene. (From Waddington and Needham.)

h.n.t. Host neural tube.
i.n.t. Induced neural tube.
imp. Implant.

Formulae of Substances Mentioned

References:

general—M. Polonowski and H. Lespagnol, *Elements de chimie organique biologique,* Paris, Masson, 1934.

O. Rosenheim and H. King, *Ann. Rev. Biochem.* (1934), *3,* 87.

sterols—C. E. Bills, *Physiol. Rev.* (1935), *15,* 1.

scymnol—R. Tschesche, *Zeitschr. f. physiol. Chem.* (1931), *203,* 263.

bile acids—H. Sobotka, *Chem. Rev.* (1934), *15,* 311.

hydrocarbons—J. W. Cook, et al., *Proc. Roy. Soc.* (B 1933), *113,* 273; (1934), *114,* 272.

aglucones—W. A. Jacobs and R. C. Elderfield, *Journ. Biol. Chem.* (1935), *108,* 497.

A Calciferol, irradiation product of ergosterol: ,vitamin D

B B-cholic acid OH COOH OH OH bile acid

C Cholesterol OH universal cellular constituent

D Scymnol OH OH OH OH conjugated as a sulphuric acid ester in elasmobranch bile

E Digitoxigenin OH OH part of the active principle of the cardiac glucoside, digitalis

F Androsterone O OH hormone of the interstitial cells of the testis. The stigmasterol of the soya bean gives another substance with male hormone properties if the side-chain is oxidised.

G Neo-ergosterol Ring II of ergosterol made completely aromatic

H *Œstrol* and *Œstrone*

hormones of the interstitial cells of the ovary

J *Luteosterone*

hormone of the corpus luteum, related to Pregnandiol

K *1,2 Benzpyrene*

L *5,6 cyclopenteno-1,2,5,6 dibenzanthracene*

oestrogenic and carcinogenic

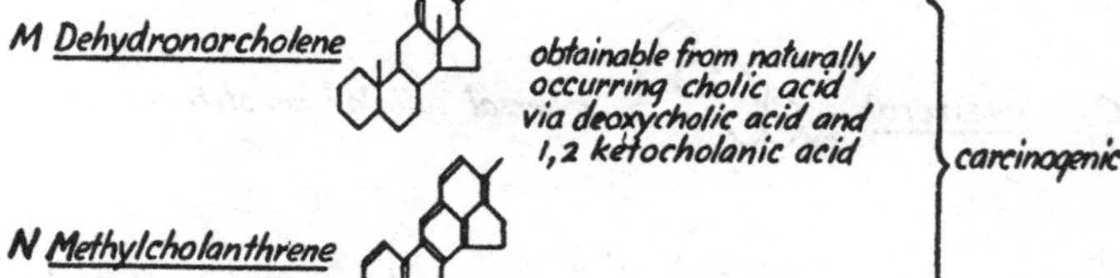

O ?

P *1,9, dimethylphenanthrene*

Q *9,10 dialkyl-9,10 dihydro-9.10 dihydroxy-1,2,5,6 dibenzanthracene*

OH R

R OH

oestrogenic and evocatory

R *1,2,5,6 dibenzanthracene*

substance may be, it is certain that its action can be imitated by two hydrocarbons which are also oestrogenic, namely, 1, 9 dimethylphenanthrene (P) and 9, 10, dialkyldihydroxy-dibenzanthracene (Q). It thus overlaps with the oestrogenic zone, and this in itself is suggestive evidence that the naturally occurring organiser has a sterol-like structure.[70] There may also be an overlap with the carcinogenic zone.

It would be almost too much to hope that any good case of a connection between all these zones should exist already in nature. Yet certain experiments of Witschi[71] suggest that there may be something of the sort in amphibian eggs kept for various lengths of

70. There may, however, be a pitfall in the argument that the naturally occurring Organiser is a sterol because polycyclic hydrocarbons can bring about the same effect. The researches of the Dutch school have shown that the hormone of plant growth, Auxin, which has the effect of lengthening cellulose cell-walls, has the structure:

COOH
OH OH OH

But the same effect is produced by another substance, Heteroauxin; isolated from human urine and from yeast, which has the formula:

COOH
N

i.e., β-indol-acetic acid. Had the hetero-auxin effect been discovered first, it would have been easy to conclude that the auxin naturally occurring in the higher plants was an indol derivative. However, it must be remembered that the solubility properties of the Organiser are strongly in favour of a sterolic nature.

71. Witschi, E., *Verh. naturf. Ges. Basel, 34* (1922), 33; *Proc. Soc. Exp. Biol. and Med., 27* (1930), 475, *31* (1934), 419.

time before fertilisation. Those which are kept a short length of time are abnormal in their sex-determining mechanism and give an abnormal sex-ratio. Those which are kept rather longer produce many double monsters, indicating a disturbance in the normal processes of production or localisation of the organiser. Those which are kept still longer produce teratomatous proliferations, whose malignant nature was demonstrated by the formation of metastases on transplantation to older larvæ. All these effects might conceivably be explained by a progressive disturbance of the normal course of sterol metabolism in these late-fertilised eggs.

There opens out before us, then, the possibility of a chemical analysis of the multitude of organiser effects which are to be found in the development of the embryo. They range all the way from cases where a hormone of the classical kind is involved to those in which a "contact" event is usually necessary. Of the former kind we may mention at random the part played by thyroxin in amphibian metamorphosis (Gudernatsch, Huxley)—a process too well known for discussion here—and the now well-proved action of a hormone produced in the head for the moulting of insects in the larval stages (Kopeć, Wigglesworth). Of the latter kind there are a great number; we may mention the induction of the lens by the eye-cup and the conjunctiva by the lens; the induction of the cartilaginous auditory capsule by the primary auditory vesicle, and of the neural by the epithelial part of the pituitary gland. Then there are the perforations of the operculum induced by the degenerating gills, and of the mouth by the fore-gut. These instances are all from amphibian development, but from the chick we may select the action of the heart in causing liver to form from the endodermal gut floor. We are as yet completely in the dark regarding the chemical nature of these organisers of the second and third grades. In some cases, such as the degenerating gills, it seems likely that special products of autolysis are responsible. The most probable prediction is that the secondary organisers are not very closely akin chemically to the primary organiser.

We are left, then, with the conception of a sterol-like substance being liberated at a given point in the developing system and "radiating" its organising power from that spot. We do not yet know how far there is an actual concentration gradient of the organiser hormone, or how far it acts by polarising and orienting other molecules strongly in its immediate neighbourhood, less strongly further away. We shall see later that there is a special significance in its sterol-like nature; the cholesteric type of paracrystal has the greatest variety of complex properties, and the protein chains no doubt need some positive polar configuration to orient them.[72] But of course much

72. N. Rashevsky, in one of his theoretical thermodynamical papers, is feeling toward such a conception (*Journ. Gen. Physiol.*, *15* [1931], 289). He is deriving expressions for the forces that will act on liquid drops suspended in another liquid, if the latter contains in solution a substance whose concentration is not uniform. These forces may exceed the electrical forces resulting from the charges on the drops, which will exert mutual attraction or repulsion.

mystery remains. If the organiser itself is of a relatively simple chemical nature, all the more morphogenetic *onus* is thrown on the competent ectoderm which reacts to it. How is it that this can fold itself into a tube? How is it that this tube can acquire a difference between its two ends?

Consideration of this problem has led to a separation of the organiser-concept into two. In the determination of the main axis of the animal by an organiser, two processes, or two aspects of the same process, can be distinguished. Roughly speaking, there are (1) the determination that an embryonic axis shall be developed, and (2) the determination of the character of that axis, i.e., that one end shall be head and the other end tail. It has been suggested that the former process be called Evocation,[73] and the latter Individuation. In a transplantation, the former process, evocation, is always performed by the graft; the latter process, individuation, seems always to be performed by the host and the graft working together, either in a co-operative or in an antagonistic manner. The former process may be regarded as strictly hormonic in character, the latter requires the intervention of a new concept, the last which we shall here discuss, namely, that of the "biological field." We do not yet know what an evoked axis is like that has never been individuated. It might be a neural

73. It is interesting to find that the more chemical the Organiser becomes, the more reluctant some biologists are to admit its importance. Aristotelianism dies hard, cf. Koehler, O., *Das Ganzheitsproblem in der Biologie* (Halle, Niemeyer, 1933), p. 183.

tube accompanied by notochord and somites in their proper relative positions, but "sausage-like," with neither head nor tail; it might on the other hand be an arrangement with a quite abnormal cross-section; or it might be a disordered mass of histologically differentiated cells. Of these possibilities, the first is the most probable. It is not likely that individuation can be performed by dead organisers or organiser extracts, but evocation, as we have seen, certainly can.

The conviction that the magnetic field has a part to play in guiding our ideas about morphological patterns is now of some seniority.[74] I pointed out

74. As a matter of fact it goes back to light-mysticism of Persian and Jewish origin. The Kabbalah, which was studied by all seventeenth-century biologists, led in one case to the extremely interesting work of Marcus Marci, of Kronland, a Bohemian. His *Idearum operatricium Idea,* published in 1635, was a mixture of purely scientific contributions to optics, and speculative theories about embryology. Thus he explained the production of manifold complexity from the seed in generation by an analogy with lenses, which will produce complicated beams from simple light sources. The formative force radiates from the geometrical centre of the embryonic body, creating complexity but losing nothing of its own power. Monsters originate from accidental doubling of the radiating centre, or from abnormal reflections or refractions at the periphery (cf. mirror reduplications, accessory organisers, etc.). Marcus Marci, in his forgotten work, thus links together the following trends of thought (a) the old Aristotelian theory of seed and blood, (b) the new Cartesian rationalistic mathematical attitude to generation, (c) the new experimental approach, in his work on optics, (d) the cabalistic mysticism of light as the fountain and origin of all things; finally, by his brilliant guess of centres of radiant energy (e) he anticipates the field theory of modern embryology. I am indebted for my acquaintance with Marci to my friend, Dr. W. Pagel; cf. my *A History of Embryology* (Cambridge, 1934).

The only parallel to this occurs, it seems, in a quarter far removed from Marci at Prag, but equally devoid of influence upon

earlier that it fulfils the criterion of divisibility of pattern which was needed for embryology (p. 70). Now if an analogy is a true one, the course of scientific thought will be greatly facilitated; thus Wrinch[75] has referred to the fundamental progress in electrostatics and current electricity due to Clerk Maxwell's genius in seeing formal resemblances between the conditions to be satisfied in those subjects and those satisfied in already solved problems belonging to other subjects. The masterly apparatus developed by Laplace and Poisson in the treatment of elasticity became available in this way for electrical theory. Can we then obtain knowledge of biological phe-

contemporary thought, namely, the *De Motu Animalium* of Borelli, the founder of the iatromathematical school. Chiarugi (*Monit. Zool. Ital., 40* [1929], 146) gives an account of the chapter on generation (Pt. II, ch. xiv). Its interest is that Borelli compares the semen to a magnet arranging iron particles in a field of force. In Harvey, too, a reference to the magnetic field can be found. In the discourse on Conception (1653, p. 539) he says: "The Woman or Female doth seem, after the spermatical contact in Coition, to be affected in the same manner, and to be rendered prolifical, by no sensible corporeal Agent, as the Iron touched by the Loadstone, is presently endowed with the virtue of the Loadstone, and doth draw other iron-bodies unto it."

In the eighteenth century, of course, traces of this are easier to find. Thus from Bourguet, one of the saner ovistic preformationists, may be quoted: "Le méchanisme Organique" which works in generation "n'est autre chose que la Combinaison du Mouvement d'une infinité de Molécules étheriennes, aëriennes, aqueuses, oléagineuses, salines, terrestres, etc. accomodées à des systèmes particuliers, determinés dès le commencement par la Sagesse suprème, et unis chacun à une Activité ou Monade singulière et dominante, a laquelle *celles qui entrent dans son système sont subordonnées.*" (Italics mine.) *Lettres Philosophiques,* etc. (Amsterdam, 1729).

75. Wrinch, D., *Proc. Aristot. Soc., 22* (1922), 208; *29* (1929), 112.

nomena by means of our knowledge of magnetic phenomena? In other words, is there an identity of formal characters in the two cases? Many, especially in recent times, have thought so, notably Gurwitsch,[76] Rudy,[77] and Weiss.[78] Schultz[79] has caustically remarked that fields without centres, without measurable field-strengths, and with no obvious lines of force cannot invite comparison with the fields of physics. But this criticism is not justified, for there was clearly a time in the history of physics when certain phenomena of magnetic fields were known although no method was available for measuring field strength.[80] The concept of "biological field" does give a powerful aid to the codification of the "*Gestaltungsgesetze*," the rules of morphogenetic order. As complex components they must first be ordered before they can be analysed.[81]

The field concept, however, has suffered greatly hitherto by an insufficient accuracy of definition,

76. Gurwitsch, A., *Archiv. f. Entwicklungsmech.*, *52* (1921), 383; and *112* (1927), 433.

77. Rudy, H., *Die biologische Feldtheorie* (Berlin, Bornträger, 1931).

78. Weiss, P., *Zeitschr. f. indukt. Abstamm. u. Vererb. lehre,* 1928, 1567, also *Morphodynamik* (Berlin, Bornträger, 1926); also *Entwicklungsphysiologie d. Tiere* (Dresden, Steinkopf, 1930); also articles in *Jahresber. u. d. ges. Physiol.*, 1922, 65; 1924, 77; 1926, 107; 1928, 70.

79. Schultz, J., *Die Maschinentheorie d. Lebens* (Berlin, Bornträger, 1929); *see also* Bertalanffy's "Tatsachen & Theorien d. Formbildung," *Erkenntnis, 1* (1930), 361.

80. e.g., in Gilbert's time.

81. It should be noted that there is an intimate relation between the field-concept and the theory of transformations of D'Arcy Thompson. This would be more generally realised if the Thomp-

leading in particular to an uncertainty in the attribution of activity to the fields. For some the fields are purely descriptive and symbolic, performing nothing, but permitting us to use a shorthand notation, as it were, for morphological form—for others, the fields are actively arranging entities, performing morphogenetic work. It is of the first importance, as Waddington shows, that this confusion should be cleared up.[82]

For Gurwitsch in the main the fields are geometrical symbolism, yet from time to time he speaks of their having an action or an influence.[83] For Weiss the fields are more active.[84] He draws three main conclusions about them. Firstly, if a certain amount of material is removed from the domain of a field, the remainder of the field manifests in due course the same pattern which it would normally have given in larger size (equipotential sea-urchin eggs, regeneration of amphibian limbs). Secondly, if unorganised but organisable material is introduced into the field domain, it is incorporated in it (regeneration of

sonian method had been applied to the change of form during ontogenesis. That particles do tend to orient themselves on progressively distorted co-ordinates is one of the most essential facts expressed by the term "field."

82. Waddington, C. H., "Morphogenetic Fields," *Sci. Progr.*, *29* (1934), 336.

83. His definition is: "Als Feld wird hier ein Raumbezirk verstanden, in welchem durch die Angabe der Coordinaten jedes beliebigen Punktes auch die Gesamtheit der Einwirkungen auf ein in betreffenden Punkte befindliches Objekt in eindeutiger Weise festgesetzt wird." *Loc. cit.*, first paper in *Archiv. f. Entwicklungsmech.*, p. 392. Rudy repeats this.

84. Cf. his definitions on p. 24 of his *Morphodynamik*.

limbs, avian development). Thirdly, two or more fields can fuse to form a larger one (fusion experiments on invertebrate eggs). A similar position is taken by Waddington, largely as the results of his experiments on induction in the chick embryo. The "individuation-field" is a term expressing the tendency of an organiser to rearrange the regional structure both of itself and of any tissue lying near it in such a way as to make that tissue part of a complete embryo.[85] In the chick, for example, a piece of neural plate in the process of being induced in ectoderm lying somewhere within the hosts' individuation-field, may follow one of three paths (1) it may become fused with the host and the two built up into a composite neural plate, (2) it may develop as a separate structure, sharing the regional character of the contiguous parts of the host, (3) its neuralisation may be suppressed, and it may be worked into the host in the form of body ectoderm. The first and second paths are the most usual.

But the "*Denkarbeit*" which has so far been devoted to the concepts of field and gradient is insufficient to enable them to bear the load of facts which, like some steel scaffolding, they are asked to carry. The gradient concept is of the two the less fundamental, for a system of gradients radiating from a

85. Waddington, C. H. and Schmidt, G. A., "Induction by Heteroplastic Grafts of the Primitive Streak in Birds." *Archiv. f. Entwicklungsmech.*, *128* (1933), 522. Cf. the idea of "Koordinative Einheitsleistung," which F. E. Lehmann develops in describing the formation of the normal amphibian neural plate (*Biol. Centrlbl.*, *53* [1933]).

centre would constitute a field, and probably no gradient exists only in two spatial dimensions. In this conceptual structure, then, we find that the term "field" is applied to at least two quite distinct things.

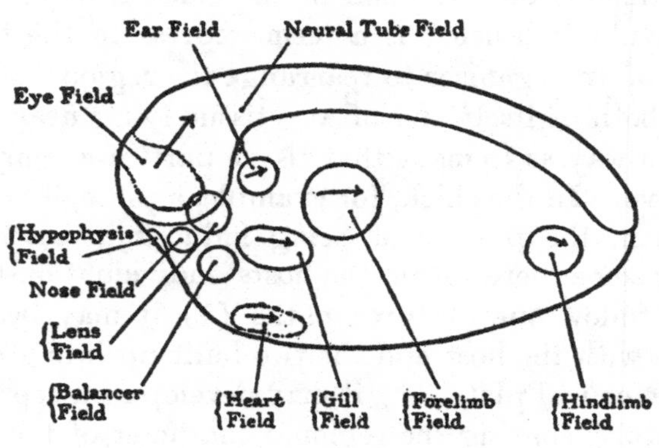

FIG. 30.

Diagram of an amphibian neurula to show the approximate localisation of the main "regional fields" so far discovered by experimental analysis. The arrows indicate that the fields are known to be polarised from their first appearance. (From Huxley and de Beer.)

In the first place, it is applied to morphological and pre-morphological domains of the character just described in Waddington's words: i.e., wholes actively organising themselves. Here it may not be illegitimate to use such expressions as "the field is a region throughout which some agency is *at work* in a co-ordinated way,"[86] or "the individuation-field *sees to it* that"[87] something or other happens. But, in the

86. Huxley and de Beer, p. 276. 87. *Idem,* p. 319.

second place, there is another kind of region also to be borne in mind, i.e., regions "possessing a general determination for the production of certain structures and undergoing progressive regional specification of detail." These, since they only imply *location*, should be more properly called "areas" or "districts."[88] Fig. 30 shows the approximate localisation of these main regional districts on the amphibian neurula. These districts represent the parts of the embryo where the presumptive rudiments of eye, limb, etc., may become determined. Under no circumstance can a limb be produced from an eye-district, but the outer edges of the districts are indistinct, and they may slightly overlap with one another. Here the concept of field-activity does not appear to be justified, although when the limb, for instance, has begun to grow out, it will constitute a true field, or a portion of the whole individuation-field, since indifferent material added to it will be organised into its own form.[89]

The districts represent, in fact, a rather different form of order. Organ-forming potency falls off from their centres in a definite measurable way. Thus Harrison's work on the limb-buds of amphibia will permit of graphical formulation.[90] The percentage of posi-

88. Waddington, C. H., *loc. cit., Sci. Progr.* (1934).

89. One of the most remarkable results of Weiss's work on the regeneration of limbs is the smallness of the pieces of tissue which can carry the field, e.g., tiny discs of arm attached to legs produce arms, not legs, on regeneration.

90. Harrison, R. G., "Experiments on the Development of the Forelimb of *Amblystoma* as a Self-differentiating Equipotential System." *Journ. Exp. Zool.*, *25* (1918), 413.

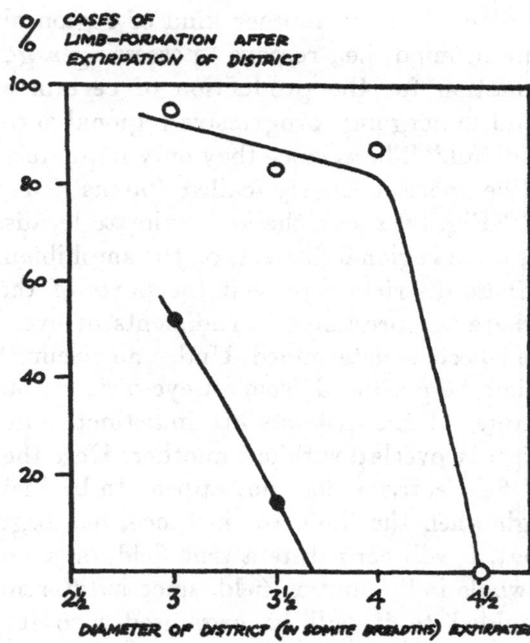

FIG. 31.
Graph showing the frequency of limb-formation after extirpation of varying disc-areas measured in somite-breadths (plotted from the data of Harrison). A way of finding the boundaries or contours of a "regional field" or "district."

tive cases of organ-formation after the excision of different sized circular pieces can be plotted against the diameter of the pieces (measured in somite-breadth), i.e., the distance of their circumferences from the presumed centre of the limb-district (Fig. 31). In such a case we are really measuring degrees of

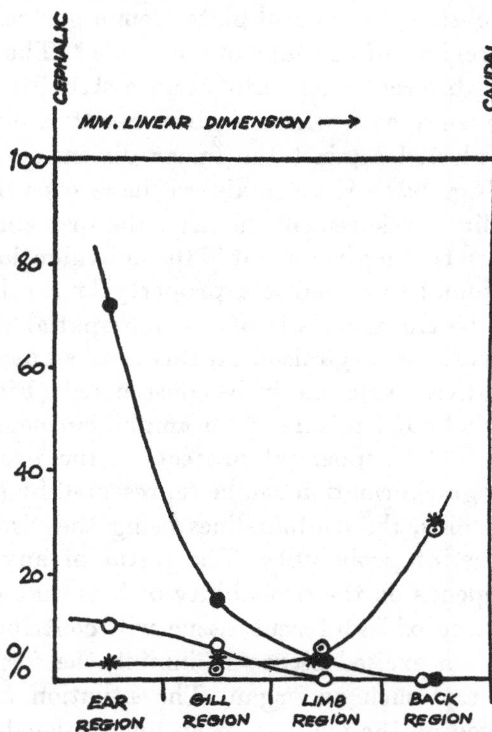

FIG. 32.

Graph showing the frequency of the formation of various structures from indifferent material implanted into different regions or districts (plotted from the data of Holtfreter).

"limb-probability," and could erect probability contours which would fix the configuration of the district round its centre. Similarly, Holtfreter has studied the results of grafting portions of presumptive epidermis or presumptive neural plate from a gastrula into various regions of the flank of a neurula.[91] The grafts undergo differentiation into various structures, and the frequency with which a given result is obtained can be plotted against the linear dimension of the body (Fig. 32). Here again we have essentially a probability measurement, namely, the probability of prediction that a given point in the individuation-field will be found to actualise a property known beforehand to be characteristic of a given spatial part of the individuated organism. On this basis a qualitative mathematical model might be constructed (Fig. 33). If the whole of one side of an amphibian neurula is mapped in hemispherical projection, the probabilities of organ-formation can be represented by a landscape of hills, the contour-lines being the "isobars," as it were, of probability. The status of any given point depends on the probability of bets that (a) an added piece of indifferent tissue will contribute to, (b) that an excised piece will inhibit, the formation of such and such an organ. The situation may be understood on the analogy of an uncompleted building—this, we say, is going to be the dining-room, but we may be quite wrong, because it might turn out to be an adjacent part, say the kitchen. There must, of course, be some underlying factor, which we could call "limb-forming intensity." In trying to picture what this can be, it would be tempting to make a

91. Holtfreter, J., "Der Einfluss d. Wirtsalter u. Verschieden Organbezirken auf die Differenzierung von angelagertem Gastrula-ektoderm," *Archiv. f. Entwicklungsmech.*, *127* (1933), 620.

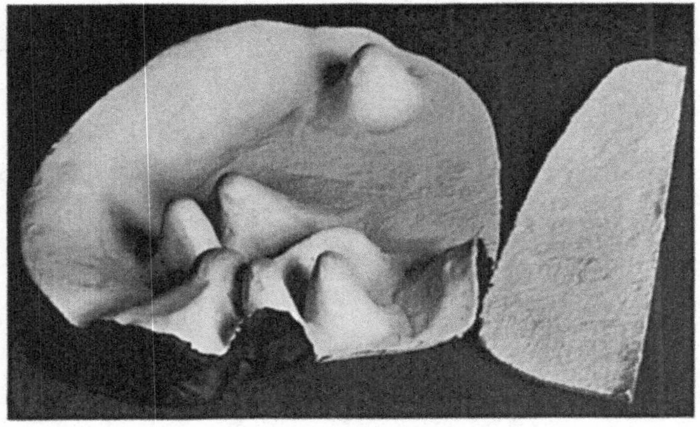

Fig. 33.

Hemispherical projection of the neurula. The two halves are united at the anterior part of the mid-ventral line. The left-hand surface is modelled topographically with probability as the vertical axis. Regional fields or districts thus appear as elevations the contour-lines of which define the probability of the fate of indifferent material implanted at any given point. The districts may be identified by comparison of this model with the two-dimensional diagram of Fig. 30. The model is seen from its anterior end.

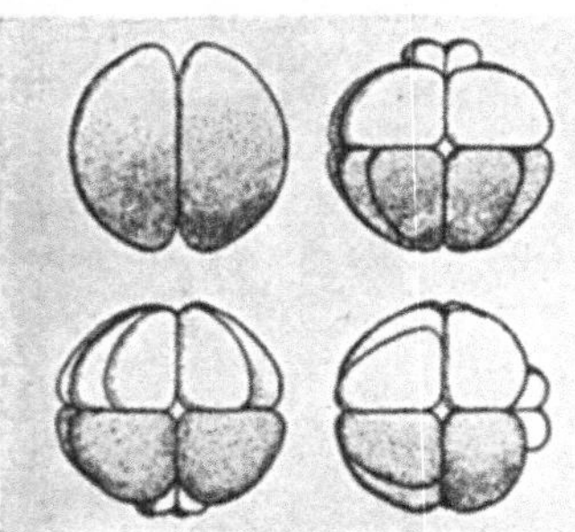

Fig. 37.

Persistence of the primary axis in the sea-urchin egg (*Arbacia*) in spite of the rearrangement of visible substances in the cytoplasm. After centrifuging, the egg becomes stratified with fat at the centrifugal pole, then clear cytoplasm, then yolk-granules with increasing amounts of pigment. The first cleavage (top left) is always at right angles to the stratification, but the micromeres are always formed at the vegetal end of the original axis, whether this coincides with the centrifugal pole of the centrifuged egg (top right), its centripetal pole (bottom left), or its side (bottom right). (From Morgan, redrawn by Huxley and De Beer.)

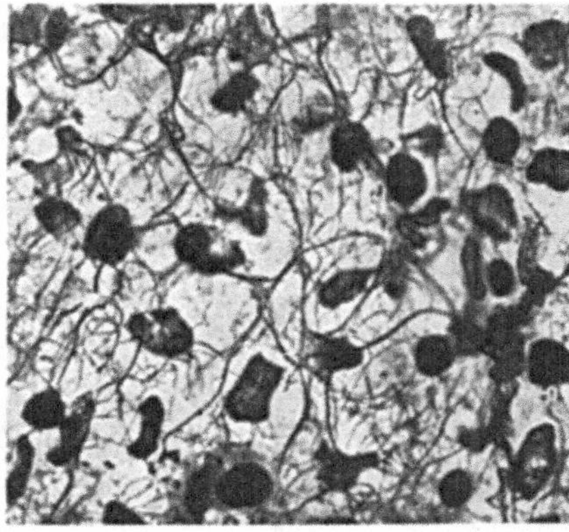

Fig. 38.

Connective-tissue fibres and fibrils in the outer wall of the exocœlom of the chick embryo at 72 hours' incubation. (From Snessarev.)

parallel between these concentric zones of organ-probability and the concentric zones of increasing randomness (probability of molecular orientation) which later on we shall consider as surrounding a central polarising molecule or paracrystalline molecular aggregate.

The situation is, of course, not so simple as it has been made to appear in the preceding paragraph. In respect of any given organ, for instance, there are to be considered (1) the district from which it normally develops (the presumptive district), (2) the district from which it will develop if the presumptive district and the surrounding tissue is extirpated, (3) the regions isolated parts of which will develop into the organ if transplanted to its district, (4) the region from which the organ can be formed by a specific or non-specific stimulus, etc. We need a vocabulary for these entities.[92]

Another, possibly fruitful, way of distinguishing between fields and districts may be derived from a remark of Weiss.[93] At a given point on a vector in a field, a field force possesses (a) a given quality, (b) a given direction, (c) a given intensity. In a field proper, such as the individuation-field, we could say that every point will possess a different value for each of these three factors, but in a district only the two latter ones will vary; the first one will be the same throughout, e.g., limb or heart. Weiss himself, however, did not distinguish between fields and districts.

92. Waddington, *loc. cit.*, *Sci. Progr.*
93. *Morphodynamik*, p. 26.

The best definition of "field" so far given is that of Waddington.[94] A field is a system of order such that the position taken up by unstable entities in one portion of the system bears a definite relation to the position taken up by unstable entities in other portions. It is, in fact, their equilibrium positions which together constitute the field effect.[95] This is in close analogy with the equilibrium reached by a particle of iron as it orients itself in a line of force of a magnet. Field laws, therefore, are simply a description of how stable configurations are reached in development; the further problem of why these and no other configurations are stable, lies beyond. But against any who wished to urge that the individuation-field is only yet another fig-leaf for our ignorance, it should be said that a dynamic description of the events is an enormous improvement upon the static descriptions of anatomy. This, too, is the answer to the question whether a field should be regarded as active or passive; it is a dynamic description of a spatio-temporal activity, not a mere geometrical picture of a momentary time-slice in the organism's history.[96]

94. *Sci. Progr., loc. cit.*

95. We could represent this by the analogy of a company of persons dressed in different colours, who are asked to walk into an arena and take up predetermined positions there. To their surprise they find they have formed the stars and stripes or the hammer and sickle.

96. After these lectures had been written and delivered, there appeared a short monograph by Koltzov, N., *Physiologie du Développement et Génetique* (Paris, Herman, 1935). Many subjects discussed in this and the following lecture appear also in Koltzov's book, and it is a pleasure to record that the conclusions arrived at are the same.

III

THE HIERARCHICAL CONTINUITY OF BIOLOGICAL ORDER

THE concept of organism, as has recently been emphasized, necessarily demands a number of subsidiary notions, such as "organic whole," "organic part," and "organising relation." "It is easy enough," writes Woodger, "to see 'intuitively' what is meant by these terms, and there has been from time to time a good deal of vague talk about 'the whole being more than the sum of its parts' etc., the difficulty is to make these notions precise in order to enable us to see how we can use them for scientific purposes. Intuition is the indispensable cutting edge of intellectual inquiry, but the ground won is not consolidated until it has passed from the stage of intuitive apprehension to that of logical analysis."[1] Until recently such logical analysis lacked the linguistic or notational technique requisite for it, and discussions of such difficult problems as the relation between embryology and genetics in the old conceptual terms could justifiably be described as the attempt to conduct a modern surgical operation with a pair of nail scissors and a potato-peeler.[2]

1. Woodger, J. H., "The Concept of Organism," *loc. cit.*, p. 8.
2. Cf. especially the forthcoming book by Woodger, J. H., *The Axiomatic Method in Biology,* in which the logistic analysis of "Principia Mathematica" is applied to biological problems.

The ordinary division of a living organism into parts composed of cells, cells themselves, parts of cells, colloidal aggregates, molecules, and atoms, has been brought by Woodger under the notion of "hierarchical order." From the present point of view we are mainly interested in "spatial hierarchy." The higher or coarser levels correspond of course to the domain of morphology and anatomy as ordinarily understood, the intermediate levels to histology and cytology, the lower or finer levels to biochemistry. On this view, structure, even, if you like, morphology, should be found fully within the sphere of biochemistry, and that this is so, the whole realm of permutations and combinations of the carbon atom abundantly illustrates. The plans of living systems, it has been said, are written in chemical molecular structure, a language with a great wealth of expression, as the latest edition of Beilstein shows.[3] The contribution of biochemistry to biology is too often thought of as if it were primarily concerned with the reactions of simple substances in homogeneous media, and the complexity of chemical structure is forgotten. Unfortunately, not only do we have to deal with extremely complicated molecules, we also have to deal with them in extremely complicated situations, that is to say, in the colloidal milieu of the living cell. When colloid chemistry is as far advanced as organic chemistry is to-day, and not till then, we shall be in a better position to follow through the course of that

3. Watson, D. L., "Biological Organisation," *Quart. Rev. Biol.*, 6 (1931), 143.

organisation whose upper and lower ends are alone visible to us now.

The spatial hierarchy is, of course, an abstraction resulting from the separation of space from time. When we take time into consideration we meet at once with other systems, including two hierarchies and one series, namely, "division-hierarchies," the hierarchically ordered systems of cells consisting of zygotes and their cell-descendants; "genetic hierarchies" the relations between division-hierarchies which result from the formation of the zygote of one by the union of gametes of others; and the series of slices of life-histories, i.e., series of which "spatial hierarchies" are the terms. For the time being we shall only be concerned with the spatial hierarchy; I mention the others to indicate the applicability of Woodger's logical methods. Division-hierarchies were much in evidence during the preceding lecture.

A member of a spatial hierarchy (e.g., a protein molecule in a colloidal particle in a nucleolus in a liver-cell in a liver in a mammal) can obviously be regarded from three points of view—firstly from that of its membership of a level (R_L relation; protein molecules in general), secondly from that of its entering into the constitution of a member of the next highest level (R_H relation; the colloidal particle), thirdly from that of its analysis into an assemblage of members of a lower level (R_A relation; its constituent amino-acids). Now it has been customary to regard an organic part almost exclusively from the third point of view. This is a natural consequence of the fact that the historically traditional method of morphology has been dissection (ἀνα, τέμνω), and correspondingly that the most widely used biochemical method has been that of analysis. We still tend to think of anatomy as Vesalius knew it, and of biochemistry as it was in the time of Liebig. But the former knew nothing of our wax reconstructions, and the latter would be amazed at the synthesis of polypeptides, hormones (such as thyroxin), or vitamins (such as ascorbic acid). As time goes on, biology employs more and more the methods of actual or conceptual synthesis, and this must be so, for it is a recognition (though not commonly a conscious one) that the R_H relations, i.e., the relations of members to the levels above them in the hierarchy are just as important as their relations to their components below. It is much to be wished that words more suitable than "above and below" or "higher and lower" could be found to indicate the levels of the spatial hierarchy, for these have honorific undertones, and it was not without intent that I referred just now to the levels as "coarser and finer." This, of course, merely reverses the psychological odium; what we need are perfectly neutral terms. Ultimately the concept we are dealing with is that of spatial dimension, not that size by itself has any importance, but because large things can contain smaller things, and hierarchical order is analogous to group-theory, with its mathematical envelopes.

There are indeed whole branches of mathematics

dealing with the assessment of complexity, which no one has the ability or the imagination to make use of for grasping the biological situation.[4] The theory of sets of points, for example, is alternatively a theory of packing. A prodigious complexity emerges from the consideration of a "hierarchical order" of ten levels of shells or envelopes, and this is assuming they are all regularly spherical. If they are annular or possess some more complicated form of connectivity, the position is still worse. Something of this kind must hold good in the actual systems of biology. The difficulty about applying topological analysis to embryonic development, for instance, is that morphological form may change considerably, as in the invagination of a gastrula, while its topological status remains unchanged. However, this is a digression; what I am concerned to point out here is that the relation between the large and the small in a biologi-

4. Thus Analysis Situs or Topology is defined as follows: "die Lehre von den Gesetzen des Zusammenhangs, der gegenseitigen Lage und der Aufeinanderfolge von Punkten, Linien, Flächen, Körpern, und ihren Teilen oder ihre Aggregaten im Raume, abgesehen von den Mass- und Grössen-verhältnissen" (Listing, 1847), or "der Aufstellung aller derjenigen Eigenschaften räumlicher Gebilde, die sich invariant verhalten gegenüber der Gruppe aller stetigen Transformationen des Raumes" (Klein, 1872). *See* Dehn, M., and Heegaard, P., *Encyclopädie d. Math. Wissenschaften,* 1907, Bd. III, Teil 1, Hälfte 1, Heft 3, pp. 153–220; and 1929, Bd. III, Teil 1, Hälfte 2, Heft 13, pp. 141–237. The morphologist will appreciate the analogy contained in the mathematical formulation of all possible knots (*see* Tait, P. G., *Scientific Papers,* Cambridge, I [1898], 273 ff.). Topology has already played an important part in organic chemistry by settling the possible number of isomers of new compounds (*see* footnote on p. 175 of Dehn and Heegaard above); this is in a sense already an application to morphology.

cal organism is a relation of continuity, in the sense that there is a morphology of atomic participation in the whole, and that there is a biochemistry of those molecular populations we call organs and organisms. The nature of the organising relations at the molecular level is presumably sufficiently defined by the concepts of affinity, valence, etc., which are used in classical organic chemistry (though this is still an assumption), and at the colloidal level there are the relations of surface force, charge, adsorption, etc., about which we know a good deal less. The crux of the problem lies in determining the nature of the organising relations at the coarser or higher levels, e.g., the nucleo-cytoplasmic relation, and those relations whose broad outlines we chart in comparative anatomy.

Woodger has here a not unimportant distinction.[5] The living organism is, he says, not exhausted by its spatial hierarchy; it contains within it certain inert entities which stand outside this hierarchy. The members of the various levels are then called "components" and the inert entities "constituents." Cartilage or bone matrix, connective tissue fibres, blood-plasma, yolk- and secretion-granules, are given as illustrations of typical constituents which stand outside the hierarchical order. Woodger goes on to say that it is in relation to constituents as opposed to components that the notion of "chemical substance" can be most safely applied, for this notion, in his view, takes no

5. *Loc. cit.*, p. 457.

account of the coarser levels of organisation, but applies to entities in which the spatial distribution of properties is uniform, so that one spatial part can be regarded as a fair sample, i.e., representative, of the whole. It is doubtful whether this distinction is of much heuristic value, questionable whether we yet know enough about the economy of the cell to be certain that such and such a part, e.g., a glycogen-granule, is merely a constituent. Indeed, from the point of view of metabolism, there are few, if any, parts of the body which can ever be said to be completely removed, *ausgeschaltet*, from possible participation in metabolic processes.[6] The breakdown of glycogen, the resorption of bone, etc., are cases in point. Much vague theorising (Jickeli[7]) and some valuable work (e.g., Terroine[8]) have been associated in the past

6. This point was well put by Hopkins in his British Association address at Birmingham in 1913 (*The Dynamic Side of Biochemistry*, p. 12). "We know that the equilibrium of a polyphasic system is independent of the mass of any one of its phases, but I am inclined to the bold statement that the integrity of metabolic life of a liver cell is as much dependent on the co-existence of metaplasmic glycogen, however small in amount, as upon the co-existence of the nuclear material itself; so in other cells, if not upon glycogen, at least upon other metaplasmic constituents." The extremest case is the interesting phenomenon of Refection in rats, discovered independently by Fridericia and by Watchorn (*Journ. Hyg.*, 1927), where vitamin B-deficient rats maintain their stores of the vitamin by systematically eating their own fæces. Here waste products, even though spatially separated, are in a sense not finally outside the metabolic machinery.

7. Jickeli, C. F., *Die Unvollkommenheit des Stoffwechsels und die Tendenz zur Stabilität als Grundprinzipien für Vergehen und Werden im Kampf ums Dasein.* (Berlin, Friedländer, 1924.)

8. Terroine, E. F., and Zunz, E., *Le Métabolisme de Base* (Paris, Presses Universitaires, 1925).

with the conception of inert ballast or "paraplasmatic substances" in the cell. Woodger's distinction needs particular care in elucidating, for it might by some be interpreted as implying that, although chemical analysis had value in the case of constituents, it had none in the case of components. On the contrary, it is clear that its application is necessarily universal if the lower, finer, levels are to be understood. The value of the distinction lies in the reminder that the position of the components in their own right must not be forgotten, i.e., that chemical analysis must take account of the coarser, higher, levels. That this is universally (though perhaps unconsciously) admitted is evident from the careful way in which biochemists now distinguish (have indeed learned by bitter experience to distinguish) between, for instance, the different organs of a mammal.[9] At the same time, we cannot assume *a priori* that differences exist between members of one level within one component, and it *may* not matter whether a sample for chemical analysis is taken, for instance, from the anterior or the posterior portion of a given organ.

In this connection it is interesting to refer to the common, but methodologically dangerous, tendency of taking an entity as an example of something else, irrespective of its meaning in its own right. Thus it is desired to study the metabolism of nerve-cells. In

9. Cf. the collection made by H. A. Krebs (Oppenheimer's *Handbuch d. Biochemie d. Menschen u. d. Tiere,* 2 Aufl. Ergänzungswerk, Bd. I, 2d half, p. 863) of the values obtained by many workers for the respiratory and glycolytic intensities of various organs.

the central nervous system these are inextricably intermixed with cell-processes (axons), their terminations, the dendrites, and the junctional or synaptic tissue. In the trigeminal ganglion, however, the cells are present in relatively much "purer" state, and such a ganglion is therefore chosen "as an example of" what neurons alone do. In this way it may easily be forgotten that the trigeminal ganglion is something more than an example of a histological class; it is a morphological part with a definite and important position in the spatial hierarchy. This may confer upon it some special character not included in the view of it as a histological sample.[10]

It is to be noted that a molecule, an atom, or an electron, if it belongs to the spatial hierarchy of a living organism, will be just as much "alive"as a cell, and one which does not belong to such a spatial hierarchy will be "dead."[11] This expresses the fact that the whole requires its components (of all levels) in order to be "alive," and that the parts require the whole in order to make their particular contribution to it by virtue of which it *is* "alive." The term "living" applies to all the components in their organising relations. If this is so, we should expect to find that

10. In the case here considered, the author did not entirely forget this. "To make use of this tissue for the study of the metabolism of nerve-cells," he said, "and to draw deductions from the results as to the behaviour of nerve-cells in general, involves the assumption that the cells are similar in metabolic behaviour to others in other parts of the nervous system." This assumption he then discussed with care. (Holmes, E. G., *Biochem. Journ.*, *26* (1932), 2005.)

11. Woodger, J. H., *loc. cit.*, p. 459.

at levels which are well within the sphere of physico-chemical analysis, there are phenomena which give us glimpses of the rudiments of wholeness. And it should follow that a contribution to the unification of biochemistry and morphology can be made by a study of such phenomena.

On this view it was significant that Vlès and Gex, examining the transparent sea-urchin egg in the ultra-violet spectrometer, obtained an absorption-curve which was not typical of the proteins, although these substances undoubtedly formed by far the largest proportion of the solid matter present in the eggs.[12] Only upon cytolysis and the death of the cell did the characteristic protein curve appear—an indication that some curious structural state was present in the egg (Fig. 34). Then much of value has come from the vigorous crop of researches which followed the perfection of the micro-manipulator as an efficient tool for handling isolated single cells.[13] When Hiller[14] made experiments on amoebae with the narcotics, ethyl alcohol, chloretone, chloroform, and ether, he found that, although their action was marked on immersion, they had no action at all when micro-injected into the cytoplasm. In all concentrations chloretone increased the fluidity and streaming movements of the interior. Alcohol might cause a reversible coagulation. Still more remarkable was the result of Pol-

12. Vlès, F., and Gex, M., "Recherches sur le spectre ultra-violet de l'oeuf d'oursin et de ses constituants." *Archives de physique biol.*, *6* (1928), 255.

13. *See* Chambers, R., *loc. cit.*

14. Hiller, S., *Proc. Soc. Exp. Biol. and Med.*, *24* (1927), 427.

lack,[15] who micro-injected solutions of picric acid, and found that concentrations which definitely coagulated proteins in the test-tube had no such action

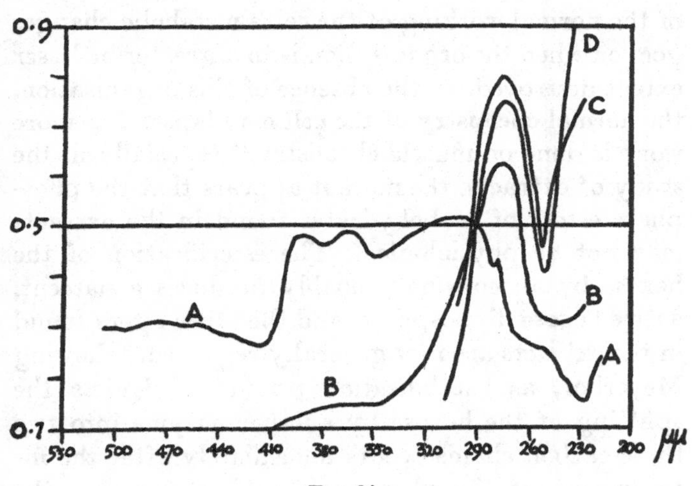

FIG. 34.

Ultra-violet spectrophotometer curves for the intact sea-urchin egg and for other solutions. (From Vlès and Gex.)
A. Normal intact egg.
B. Egg cytolysed by hypotonic solution.
C. Egg cytolysed by crushing at pH 6.7.
D. Solution of egg-albumen under comparable conditions.

in the interior of the living amoeba. Only if there was a local injury at the point of injection did the picric acid coagulate the protoplasm. In contrast to the non-toxicity of the reagent within was its extreme

15. Pollack, H., *Proc. Soc. Exp. Biol. and Med.*, *25* (1927), 145. The neutralisation of the picric acid by the base in the cell-interior probably plays a part here.

toxicity when applied to the external surface. The interest of these facts hardly requires emphasising.

In elucidating the finer levels of biological organisation it is important to discover precisely how much of the normal working of the cell's metabolic changes goes on when the organisation is to a greater or lesser extent destroyed. In the absence of this organisation, the normal chemistry of the cell may lapse. The more work is done on muscle chemistry,[16] especially in the study of extracts, the more it appears that the phosphate esters of carbohydrates found in the extracts may not be physiological. The esterification of the hexose by an enzyme probably produces a nascent, active hexose diphosphate, and the other esters found in the extracts are now generally regarded, following Meyerhof, as "stabilisation products." Unless the splitting of the hexose by another enzyme into two three-carbon chains occurs immediately after the introduction of the phosphoric acid group into the molecule, more stable products are formed. Embden's monophosphate is possibly one of these stable products, and the Harden-Young diphosphate certainly is. These products are not found in the living muscle; they do desmolyse, but more slowly than the active ester. Stabilisation, then, occurs in extracts and not in the living cell, presumably because the enzymes are there so contiguously situated, so juxtaposed, that the whole of the active monophosphate is glycolysed while still "nascent." It is difficult to picture

16. *See* Needham, D. M., *The Biochemistry of Muscle* (London, Methuen, 1932).

an integrative mechanism except in the distinctively morphological terms of contiguous situation of the two enzymes at an intracellular surface.

The phenomenon of the potential accumulation of stabilisation products is met with again in the first stage of the breakdown of the polysaccharides into lower carbohydrates. Thus a glycerol extract of muscle, which contains amylase but no glycolytic enzyme, produces a trisaccharide which cannot be attacked by the glycolytic enzyme. Under such conditions the product of the first breakdown stage is a product which cannot be further broken down. Case[17] found that if a preparation of muscle amylase was added to muscle extract, a similar trisaccharide was formed, while no lactic acid production took place. The muscle extract itself contained amylase, and the products of *its* activity could be esterified and broken down, for before the addition of the extra amylase lactic acid production was proceeding. Case's explanation was as follows: in the normal breakdown the amylase is in some way situated (e.g., on a colloidal particle) so as to be in close contact with the next enzyme in the breakdown chain, the two enzymes working almost simultaneously on the glycogen molecule. But in the amylase preparation added, this association with the esterifying enzyme had been destroyed. Then the added amylase combined with all the glycogen in the extract and turned it into a substance incapable of further breakdown, leaving the amylase-

17. Case, E. M., *Biochem. Journ.*, *25* (1931), 561.

esterification complex with nothing to do, so that lactic acid production ceased.

What is new in these conceptions is the realisation of the part played by enzyme organisation. Case points out that the myozymase complex is extremely sensitive to rise of temperature, pronounced effects being observable by even short exposures to 37°, and suggests that this is not due to enzyme destruction but rather to the dissolution of some delicate colloidal link between one part of the complex and another. "In the living muscle itself," writes D. M. Needham, "a very close co-ordination of the enzyme systems is probably arranged. In muscle extract, ester may accumulate, and in the case of glycogen as substrate there is always some accumulation of reducing carbohydrate which does not break down further. But in the muscle, resting or fatigued, there is very little reducing carbohydrate, no diphosphate, and little monophosphate."[18] Oxidising enzymes, too, have been shown to possess a "morphological" character. Cook, Haldane, and Mapson,[19] working with the respiratory catalysts of bacteria, concluded that the relation between a dehydrogenase (enzyme activating metabolites) and the oxygenase or *Atmungsferment* (enzyme activating oxygen) associated with it is probably one of intimate juxtaposition at some intracellular surface.

Many more examples might be given to exemplify

18. *Loc. cit.*, p. 45.

19. Cook, R. P., Haldane, J. B. S., and Mapson, L. W., *Biochem. Journ.*, *25* (1931), 534.

this statement. We are coming to see that an enzyme working inside a living cell is characterised not only by what it can perform when isolated, but also by such factors as accessibility, specificity, etc., which modify it *in vivo*. The classical work of Warburg and Meyerhof[20] in 1912 showed how the respiration of sea-urchin eggs can be reduced to 12 per cent of the normal rate when the cellular structure is abolished by cytolysis. More recently Penrose and Quastel[21] made a very interesting study of the effect on enzymic activity of cytolysing a bacillus. *Micrococcus lysodeikticus* is an organism which "lyses" instantly when brought into contact with the "lysozyme" of egg-white, a property which permitted of the accurate study of the fate of a number of its most important enzyme systems. It was found that the *Atmungsferment*, catalase, fumarase, and urease, showed a slightly increased activity after cytolysis, but that all the dehydrogenases (e.g., for glucose, fructose, glutamic acid, lactic acid, etc.) were completely destroyed. Peroxidase alone retained its original activity unchanged. Since the hydrogen-activators were so much affected, the general upshot was to reduce the oxygen-consumption of the bacteria by over 90 per cent.

20. Warburg, O., and Meyerhof, O., *Archiv. f. d. ges. Physiol.*, *148* (1912), 295.

21. Penrose, M., and Quastel, J. H., *Proc. Roy. Soc. B*, *107* (1930), 168. Another instance of the significance of structure in bacterial chemistry may be given. *B. coli* can use nitrate to oxidise formic acid anaerobically when intact. But after digestion with trypsin, a carrier such as pyocyanin is needed, although the two enzymes still persist in active state.

But if in these cases the destruction of the cell-organisation dissolves enzyme contiguity and destroys some essential factors, leading to an overwhelming loss of respiratory activity, the opposite effect may be found, and the respiration may increase owing to the coming together of molecules which before could not do so. Keilin,[22] in his Croonian lecture, has cited the case of a mushroom in which the bruised portion will take up many times as much oxygen as the normal remainder of the organism. And Brachet[23] has demonstrated that the gastrulæ of amphibia take up five times as much oxygen after cytolysis as they did before. Micro-morphological structure, then, may be one of the most important limiting factors governing the rates at which cell-reactions proceed. We know now of several cases in nature where this has been proved to be normally the case; I may mention two. Firstly, the exhaustive investigations of Runnström[24] have shown that the unfertilised sea-urchin egg is a system in which the *Atmungsferment* is not fully saturated with substrates, whereas after fertilisation the completest accessibility prevails and the respiratory rate is accordingly much higher. It would

22. Keilin, D., Croonian Lecture, *Proc. Roy. Soc. B* (1935).
23. Brachet, J., *Archives de Biol., 45* (1934), 611.
24. Runnström, J., *Protoplasma, 10* (1930), 106. Similarly, exposure of living plants to high oxygen pressures does not initiate certain oxidations which proceed readily in dead plants in the air. Since the living plant cells are as freely permeable to oxygen as dead cells, more oxygen may enter them at high pressure, but the reacting substances (chromogen and oxidase) are kept apart by some phase boundary as long as the cells are alive. (Harvey, E. N., *Journ. Gen. Physiol., 5* [1922], 215.)

follow that the respiration of the unfertilised egg would be but little affected by KCN and CO, while that of the fertilised egg would be greatly affected; this is actually found to be the case. The activity of the dehydrogenases, on the other hand, is much the same before and after fertilisation. The second case is that of the egg of the grasshopper, which goes through a long period of quiescence during its development, known as the diapause. In this state morphological and histological differentiation is suspended, growth at the expense of the yolk ceases, and the respiration falls to a low and constant level. The diapause may last six months. In the silkworm industry it has great commercial importance. Now Bodine and Boell[25] found that the respiration during the diapause was not inhibitable by KCN and CO, and that it was quantitatively of the same order as the uninhibitable fraction of the respiration found during active development. They therefore concluded that the oxygen-activating, cyanide-susceptible *Atmungsferment* had in some way been side-tracked or rendered inaccessible, just as in the unfertilised sea-urchin egg. The mechanism governing diapause must be able to affect this side-tracking. In the absence of oxygen activation by iron-pyrrol compounds, growth and differentiation appear to be impossible, although the basal metabolism of what has already been formed can be carried on by aerobic dehydrogenases or reversibly oxidisable-reducible pigments, possibly flavines.

25. Bodine, J. H., and Boell, E. J., *Journ. Cell. and Comp. Physiol.*, *4* (1934), 475; *5* (1934), 97.

These relations between the cellular structure and the Wieland-Thunberg and Warburg-Keilin oxidation mechanisms respectively, are of the greatest interest. But still more light is thrown on the importance of micro-morphology by the work of Warburg and Christian[26] on red blood corpuscles. If red blood corpuscles are cytolysed with water, a solution is obtained which will neither respire, nor glycolyse with glucose, nor show any glycolysis or respiration when methylene blue is added to it. But it does still retain the power of glycolysing and oxidising hexosemonophosphate (Robison) in the presence of methylene blue. Now from red blood corpuscles Warburg and Christian could prepare an enzyme and co-enzyme solution which will make hexosemonophosphate react with molecular oxygen in the extract, i.e., something which is quite beyond the capacities of the intact cell. In the intact cell some mechanism or structure must be present which permits of no greater activation of the substrates than the *Atmungsferment* can deal with. Hence KCN and CO should affect the intact cells but not the enzyme-co-enzyme solutions; this is found to be the case.

In our speculation about the micro-morphology of the cell, however, there is need for a methodological warning. It is very important that we should be on our guard against any tendency to ascribe to cell-organisation all the mysterious or obscure phenomena that we do not understand. Thus Peters' paper at the

26. Warburg, O., and Christian, W., *Biochem. Zeitschr.*, *242* (1931), 206.

Faraday Society Symposium of 1930 was followed by a spirited contribution from Woolf, who urged that the fullest exploitation should be made of the capabilities of enzyme co-operation in almost homogeneous media.[27] If, for example, an enzyme in a system of linked enzyme reactions is sensitive to one of the intermediate products produced by one of the other enzymes, a false equilibrium will be established.[28] There would thus be enough possibility of mutual control to account for the observed orderly state of affairs, without recourse to further hypothesis. It is probably a sound methodological principle that only where explanations on these lines can be shown to be incapable of accounting for the facts, more complicated organisation should be called upon.[29] But it will be noted that a chain of enzyme reactions involves heterogeneity in so far as enzymes,

27. Woolf, B., Faraday Society Symposium (London, 1930), p. 816.

28. Adenylpyrophosphate is a perfect instance of this. It is the co-enzyme of myozymase, i.e., the enzyme which converts glycogen into lactic acid, but at the same time it is reversibly broken down, as we shall see, in the muscle.

29. Tauber, H. (*Chem. Rev.*, *15* [1934], 99) distinguishes between three classes of biological catalysts:

1) specific, cell-independent, thermolabile catalysts, e.g., lipase or trypsin;
2) specific, cell-dependent, thermolabile catalysts, e.g., the urea-forming mechanism;
3) unspecific, cell-independent, thermostable catalysts, e.g., glutathione, ascorbic acid, cytochrome.

The lipase of the liver and salivary gland should, according to Bamann, be added to the second of these sections, for it can only be liberated by autolysis (Bamann, E., Mukherjee, J. N., and Vogel, L., *Zeitschr. f. physiol. Chem.*, *223* [1934], 1; *229*, 1, 15).

The exact criterion we adopt here is likely to be of some impor-

according to our present knowledge, only work on colloidal carriers, and the chain in itself *is* organisation.

All this may be illustrated from the subject of muscle chemistry, still in a very rapidly advancing state. Up to the summer of 1934 the provision of energy for the muscle contraction was believed to occur along a chain of parallel, not linked, reactions. The energy derived from the breakdown of glycogen to lactic acid was utilised in the synthesis of adenylpyrophosphate, and the energy derived from the breakdown of this compound provided in turn the energy required for building up a supply of phosphagen. This compound then on breaking down handed energy direct to the contractile machinery of the muscle. All of these reactions were found to proceed together in an orderly way in the muscle extract, with the exception of the last one. As no analogy for this transference of energy from one independent reaction to another existed in ordinary chem-

tance. If we accept inactivity of an enzyme in *brei* and activity in tissue-slice as sufficient evidence of attachment to micro-morphological structure, we may be misled, as Krebs has pointed out (Krebs, H. A., *Biochem. Journ.*, *29* [1935], 1639), by the greater dilution of *brei* conditions. A reaction involving a threefold collision will naturally be much more strongly affected by a given degree of dilution than a simpler reaction, and might consequently appear to be more dependent than the latter on the integrity of the cell-structure. Facility of removal by various methods, such as autolysis, or various treatments with different effects on the proteins, is another criterion, however. Krebs' caution will not apply to cases such as that of the desmo- and lyoglycogens (Willstätter, R. and Rohdewald, M., *Zeitschr. f. physiol. Chem.*, *225* (1934), 103.

SCHEMES OF MUSCLE METABOLISM

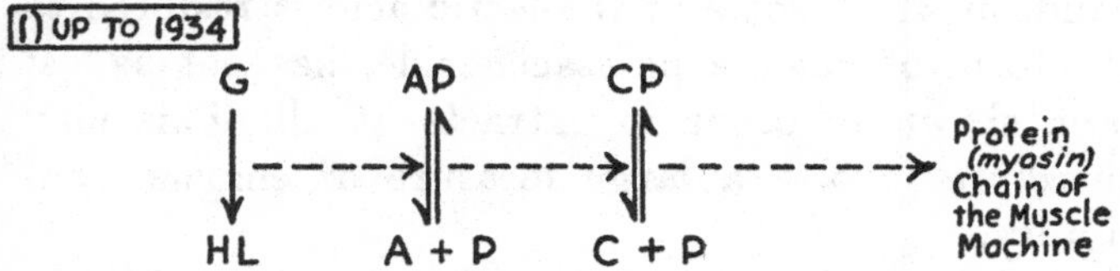

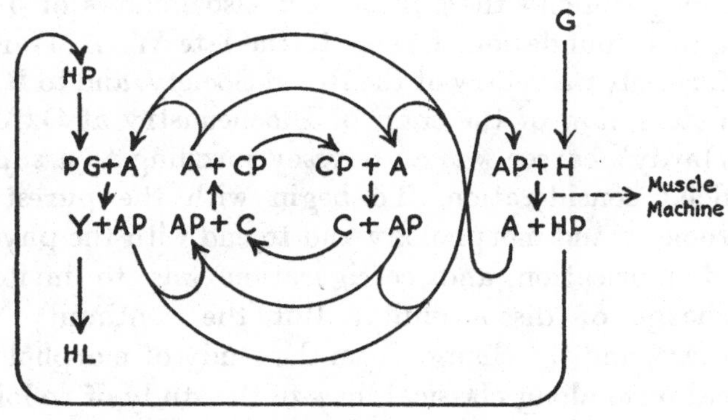

KEY

G = glycogen	C = creatine
HL = lactic acid	H = hexose
P = free phosphate	HP = hexosephosphate
AP = adenylpyrophosphate	PG = phosphoglyceric acid
A = adenylic acid	Y = pyruvic acid
CP = phosphagen	

istry, it was natural to assume that a "morphological" contiguity of enzymes on interfaces was responsible. The scheme is now, however, being replaced by another in which the reactions are chemically coupled (diagram). The necessity for the hypothesis of contiguity is therefore not so great.

The process of oxidative recovery, on the other hand, in which some of the lactic acid is restored to the form of reserve polysaccharide, has not as yet been shown to occur in extracts at all. This may therefore involve a large measure of enzyme contiguity.

By a strange coincidence, if it is a coincidence, two of the investigators who have most influenced my thought along these lines were also Fellows of John Caius' foundation. I refer to the late W. B. Hardy, formerly Secretary of the Royal Society, and to R. A. Peters, now of the chair of Biochemistry at Oxford. Hardy's career was an odyssey puzzling to a superficial consideration. To begin with the purest of zoology and morphology and to end with the physics of lubrication and refrigeration was to invite a charge of discontinuity. But the continuity was there, and the change from the study of morphological form along classical lines to the study of colloidal surfaces along lines fundamentally new and original was as logical as it could be. It arose from the conviction that the problem of biological organisation would never be solved by the mere description of form at the higher or coarser levels, but that some bridge was essential between the largest organic molecules and the smallest intracellular structures. It was precisely because the riddle of the organising relations of living matter needed new sorts of physics for its solution that Hardy the zoologist became Hardy the physicist.

His earliest papers written about 1890 show him

employing all the resources of old-fashioned comparative morphology and histology in the study of reproduction in hydroids, and phagocytosis in various invertebrates. But in 1899 came the cardinal piece of work which changed the whole of his outlook, and made a permanent impression on biology. The fluidity, delicacy, and opaqueness of living objects had in previous centuries been one of the greatest obstacles with which biologists had had to contend, so much so, indeed, that the history of the preformation controversy in the eighteenth century showed the impossibility of progress until some hardening or fixing agent could be found. Alcohol as a fixing agent for small chick embryos had been introduced by Robert Boyle in 1666, and "distilled spirits of vinegar" had been used with much effect by Antoine Maître-Jan in 1722, but it was not until the early years of the nineteenth century that alcohol and other fixatives began to be generally employed. Only then could histology begin. During the second half of the nineteenth century a great deal of work was done and much was written on the minute structure of protoplasm. It had value, but it had one supreme drawback, a quite uncritical attitude regarding the effects which the fixing agents produced.

When Hardy wrote in 1899, the living cell-protoplasm was regarded by some as being composed of two substances, one of which was disposed as a visible contractile net, a relatively rigid visible framework, or as free filaments. Others regarded protoplasm as being composed of a more solid material containing visible vacuoles filled with fluid. "I endeavour to show," said Hardy, "that the lack of consonance in the views held as to the structure of cell-protoplasm is traceable in the main to the fact that they are largely based on details of structure visible both in fresh and in fixed cells which are the result of the physical changes which the living substance undergoes in the act of dying or in the hands of fixatives." In his classical paper[30] he definitely established and made precise the view previously adumbrated by Bütschli that protoplasm has an emulsion-like or foam-like, as we should now say, colloidal, structure, and he showed that fixation produces an insoluble, distorted, rearranged modification of it. On the third page of his paper he introduces the name of Thomas Graham, and thus brings about that wedding between colloid physics and general biology which has been the distinguishing mark of the modern period. Without Hardy's nuptial blessing, the famous aphorism of Hopkins—"life is a dynamic equilibrium in a polyphasic system"—could never have been made. Hardy's conclusions are worth quoting: "A study of the action of reagents upon colloidal matter shows that when an insoluble modification is formed there is a separation of solid particles which are large molecular aggregates, and that these become linked to-

30. "On the structure of cell-protoplasm: the structure produced in a cell by fixative and post-mortem change, the structure of colloidal matter, and the mechanism of setting and coagulation." *Journ. Physiol.*, *24* (1899), 288.

gether to form a comparatively coarse solid framework having the form of an open net which holds fluid in its meshes. Reagents confer a structure on colloidal matter which differs in most cases in kind, in some cases, in degree, from the initial structure. Hence it is inferred that the structure seen in cells after fixation is due to an unknown extent to the action of the fixing reagents."

Now this basic work led in Cambridge and elsewhere to two distinct lines of investigation. In the first place, by destroying that confidence in histological preparations which the older biologists had entertained, it caused a great intensification of effort toward the elaboration of methods for examining the living cell in as natural conditions as possible. This led in the hands of Strangeways to the technique of tissue-culture,[31] and to the triumph of those who first saw living cells go through mitosis under their very eyes. But in the second place, and this was the movement with which Hardy himself was associated, it brought up the whole question of our knowledge of the colloidal state, and it plainly indicated that a large part of the secret of biological organisation lay there. For the rest of his life, diverging more and more widely from biological material, but always returning from time to time to biological theory, Hardy unravelled the regularities of the colloidal state. From "The mechanism of gelation in reversible col-

31. We shall not forget that the original birthplace of this was Yale (R. G. Harrison).

loidal systems" of 1900[32] to "Adsorption; a study of availability and accessibility" in 1932,[33] the line was uninterrupted.

In his Guthrie Lecture of 1916[34] he described the living cell as a "mixture of colloidal slimes of varying degrees of fluidity" and drew attention to the many strange phenomena deriving from its organisation. The colloid particles were to be pictured as strain centres and the continuous medium as having the properties of unannealed glass. Mechanically, he reminded his hearers, the living cell has many points of similarity to a Prince Rupert's Drop, and protozoa may disrupt explosively if the surface membrane is cut. Very important for the present theme was his contribution to the Loeb memorial volume,[35] in which he advanced a theory of trophic action which still remains, indeed, the only attempt to explain that curious phenomenon. Tonus and heat-production of muscles depend on continuity of nerve-supply, but it is not believed that the nerve fibre here acts by conveying impulses to the muscle. Hardy suggested that its effect might be due to orientation of molecules. A layer of molecules oriented by an interface might, he thought, be of the order of at least a thousand molecules in thickness. This, he wrote, would be a humble example to place beside the colossal examples fur-

32. *Proc. Roy. Soc.*, *66* (1900), 95.
33. *Idem*, *138* (1932 A), 259.
34. "Some Problems of Living Matter," *Proc. Physical. Soc.*, *28* (1916) (ii), 99.
35. "Molecular Orientation," *Journ. Gen. Physiol.*, *8* (1927), 641.

nished by living matter, "but when one has for years contemplated a scientific problem towards the explanation of which nothing could be advanced, even the slightest clue is welcome."

This suggestion derived from the work of Hardy and Nottage[36] in which it was shown that orientation effects may be transmitted from a surface such as that of a metal capable of influencing a surrounding film for a distance of 7 to 8μ. "Here we have at once," wrote Peters, commenting upon this later, "a means by which a mosaic may radiate its effects throughout the cell. It is possible to appreciate how a co-ordinate structure may be maintained even in a medium which is apparently entirely liquid. We have in effect chains of liquid crystals which need not be more than a few molecules thick. Once we have seen this, it becomes possible to develop some dim conception as to how the miracle occurs that colloid molecules, which have apparently no particular location in the cell, can yet control the synthetic and other reactions which are proceeding. This theory also is all that is needed to enable us to understand how substances can reach a special site in the cell. Between the chains of molecules, fixed by the radiating webs, there will exist paths from the external to the internal surface."[37]

Perhaps the mainsprings of Hardy's work can best

36. "Studies in Adhesion. II." *Proc. Roy. Soc., 118* (1928 A), 209.

37. Harben Lectures, see note 40.

be studied in his Chemical Society lecture of 1925[38] and his address to the Colloid Symposium of 1928.[39] Both are so distributed between biological and physical problems that it would be hard to say to which field either belonged. Why is a biologist, asked Hardy, justified in sticking so obstinately to the problem of lubrication, to the problem of the distance to which the matter on either side is modified by the field of force at an interface, in spite of the handicap of his imperfect training for the task? It is because it offers a slight but precarious foothold in the most obscure and most fundamental region of biology. Growth and regeneration, differentiation and determination, considered as physical processes, are almost wholly beyond existing knowledge. How can living matter preserve its space pattern? How can it be the seat of so many peculiar and special chemical processes, how maintain within itself sinks and sources of energy? Its multitude of suspended particles can be swept aside by the centrifuge, leaving a hyaline optical vacuum. The complexity of this appears to be the complexity neither of a machine nor of a crystal, but of a nebula. Gathered into the nebula are units relatively simple but capable by their combinations of forming a vast number of dynamical structures into which they fall as the distribution of energy varies.

I quoted a passage just now from Peters. As I write I have before me the notes which I took of his

38. *Journ. Chem. Soc.*, *127* (1925), 1207.
39. *Colloid Symposium Monographs*, *6* (1928), 7.

lectures in the Cambridge biochemical laboratory on what was called "physico-chemical cytology." It was about the time at the end of the war[40] when the work of Hardy[41] was being so brilliantly extended in the United States by Harkins[42] and by Langmuir.[43] Peters, "groping his way" as he said "towards some explanation of the action of the toxic gases,"[44] first began to appreciate the importance of Hardy's observations on the relation between interfacial tension and chemical constitution. This work showed that surface force depended on organic chemical structure. "With the difficulties of physical theories of adsorption confronting me," he wrote afterwards, "I can hardly describe now how I rejoiced when I read the work of Harkins and Langmuir."[45] This enthusiasm was shared by those who listened to Peters' lectures. They gained a new confidence in the structural symbolism of organic chemistry. Their eyes were

40. "Up to that time and for some time later, I think it is fair to say that it was orthodox to keep physical and chemical views strictly apart, as far as surfaces were concerned." (Peters, R. A., Harben Lectures, *Journ. State Med.*, *37* [1929], 1, p. 6.)

41. Hardy, W. B., "Influence of chemical constitution on interfacial tension," *Proc. Roy. Soc.*, *88* (1913 A), 304. Cf. the passage in his Guthrie Lecture, p. 111, "Before the war I had begun to observe particles of interfaces under the ultramicroscope with the object of detecting structure. It is certain that the Brownian movements are sometimes damped in a remarkable way in this region."

42. Harkins, W. D., *Journ. Amer. Chem. Soc.*, *39* (1917), 354, 541.

43. Langmuir, I., "Constitution of Solids and Liquids," *Journ. Amer. Chem. Soc.*, *38* (2) (1916), 2221; *39* (2) (1917), 1848.

44. Harben Lectures, *loc. cit.*, p. 6.

45. *Idem*, p. 7.

opened to the biological possibilities arising from the orientation of molecules at interfaces. They learned to think[46] in terms of the forces causing a palmitic acid molecule, for instance, to take up a position at right angles to the water surface, with its carboxyl group "dissolved" and its chain of methyl groups lying toward the air. "With this evidence behind us," wrote Peters, in his important Harben lectures, "we can picture with reasonable probability that a cell surface whether external or internal may be made of molecules so anchored as to constitute a chemical mosaic. This conception makes 'co-ordinative biochemistry' possible." What Peters calls co-ordinative biochemistry[47] is precisely what I mean by the extension of morphology into biochemistry and the bridging of the gulf between the so-called sciences of matter and the so-called sciences of form.

In order to show how different the chemistry of the finer levels of living organisation must be, he gives the instance of a cell on the limit of microscopic visibility, some 0.2 μ in diameter.[48] According to the laws of mass action applied to the dissociation of water, we

46. And they had the additional advantage that N. K. Adam was at that time still a member of the Cambridge biochemical laboratory, and was beginning his well-known researches on monomolecular films (Adam, N. K., *The Physics and Chemistry of Surfaces,* Oxford, Clarendon Press, 1930).

47. In the Harben Lectures and in his contribution, "Surface Structure in the Integration of Cell Activity," to the Faraday Society Symposium, *Colloid Science Applied to Biology* (London, 1930), p. 797.

48. This extreme smallness of the units of the living system may be made the basis of a theory which regards living organisms as especially constructed so as to transmit into the macroscopic world

are accustomed to suppose that 1 water molecule in every 555,000,000 is dissociated at neutrality and room temperature. Yet the volume occupied by this water would be a cube with a side of 0.25 μ length. Therefore each such cell or organism can only contain one hydrogen ion at pH 7. How then can the statistical laws of ordinary chemistry apply to the chemistry of the cell? Dissociation itself is certainly not unaffected by the presence of interfaces in the system, for Peters himself showed[49] that the dissociation constant of a fatty acid might differ by several pH units according to whether it was adsorbed at an interface or free.

Let us pause for a moment to take stock of the argument as it has so far run. A logical analysis of the concept of organism leads us to look for organising relations at all the levels, coarse and fine, of the living structure. Biochemistry and morphology should, then, blend into each other instead of existing, as they tend to do, on each side of an enigmatic barrier. The whole life work of Hardy,[50] continued in certain aspects by Peters, showed that the chemical

the effects of non-statistical phenomena occurring in minute phases. Intra-atomic types of action may have an assymetric or directive character which distinguishes them from mass actions in which a multitude of random motions cancel each other out. "The spontaneous development of a high degree of small-scale structural differentiation, the unique characteristic of protoplasmic processes as contrasted with those of non-living chemical systems, seems explicable only by reference to intra-atomic factors of determination." (Lillie, R. S., *Journ. Philos.*, *29* [1932], 477; *Amer. Nat.*, *68* [1934], 304; *Philos. of Sci.*, *1* [1934], 296.)

49. *Proc. Roy. Soc.*, *133* (1931 A), 140.

50. A complete bibliography appeared in the *Caian* (1934).

structure of molecules, the colloidal conditions in the cell, and the morphological patterns so arising, are inextricably connected. Instances have been given of the way in which organisation may appear already at the chemical level. The conception of the living cell as having as complex a set of interfaces, oriented catalysts, molecular chains, and reaction-vessels as the whole organism has of organs and other anatomical structures, comes into view.

But, it will be asked, is not this complexity somewhat reminiscent of the theory of biogen molecules? In Cambridge there has always been a strong tendency, associated with the name of Hopkins (for nearly twenty years now truly *in loco parentis* to me) to regard the biogen molecule theory as one of the grand heresies of biochemical thought.[51] This orthodoxy is both justified and intelligible. In their famous Croonian lecture[52] Fletcher and Hopkins described how the "inogen molecule" of Hermann (1867) invoked as the explosive unstable precursor of lactic acid and carbon dioxide in the muscle, and pictured as containing oxygen from the air in combination with it, was taken over by Pflüger in his studies of physiological combustion. Pflüger's giant molecule

51. Hopkins, F. G., "The Dynamic Side of Biochemistry." Address to the British Association, Birmingham, 1913; p. 11 of the reprint.

52. Fletcher, W. M. and Hopkins, F. G., Croonian Lecture, "The Respiratory Process in Muscle and the Nature of Muscular Motion." *Proc. Roy. Soc., 89* (1917 B), 444; Hopkins, F. G., Herter Lecture, "The Chemical Dynamics of Muscle," *Johns Hopkins Hospital Bulletin, 32* (1921), 359.

was Hermann's inogen "amplified and illustrated with great wealth of rhetoric, but without significant change or fresh experimental support." It was thus conceived, said Fletcher and Hopkins, that the chemical processes of life in all cells consisted essentially in the building up of elaborate, unstable, and oxygen-charged molecules into the mystical complexes of irritable protoplasm.

That the postulate of a previous inclusion of oxygen within the muscle elements was inadequate and that for inogen there was no lasting experimental support, is known to all, and may be read in these classical papers. The speculations of Pflüger upon "dead" and "living" protein were relegated to the scrap-heap. The biogen or inogen molecule, in so far as it was thought of as living, possessed, indeed, a logical status remarkably similar to that of the entelechy, discussed in the preceding lecture. It hardly did more that retell, in tabloid form, the perplexing story of the actual facts. If the cell was one giant molecule, it was hardly worth while calling it a "proteid" or inventing any other new names for it. The situation was reminiscent of Mulder's ascription of elementary formulæ to the parts of the body (1844)[53] so that skin, e.g., was said to be $(C_{40}H_{66}N_{12}O_{15})S$, though no proof of the formula was offered. In such a

53. Mulder, G. I., *Versuch einer allgemeinen physiologischen Chemie* (Braunschwig, 1844). This work is discussed in the suggestive paper of Färber, E., "Stoff und Form als Problem der biochemischen Forschung, eine geschichtliche Betrachtung," *Isis, 21* (1934), 187.

case we see the importance of the Axiom of Atomicity.[54] Entities which we believe to be intrinsically different can only be brought into the field of scientific discourse by being analysed into their constituent parts. We can then say that the intrinsic differences are due to variations in the number, nature, or relations of these parts. If the entities cannot be analysed into parts, nothing remains but to label them or name them, and the scientific method comes to a stop. This analysis was exactly what lay at the bottom of the great advance made by Fletcher and Hopkins.

But analysis does not necessarily mean the actual breaking-down of an entity. If we used the analogy of a box made of dark glass with some complicated apparatus inside it, there are obviously two ways in which we could find out about its internal parts. We could firstly break the glass and dissect the system in the traditional anatomical way, but secondly we could shine a bright light through the object in various directions, and so by increasing our powers of vision, we could obtain the same, and possibly a little more, information. This is the position we are in as regards the organisation of the living cell. The only reason why we need not be afraid to credit the cell-interior with a structure is that we have to-day at our disposal various methods of increasing the acuity of our vision, according to the analogy just made.

Of the new means of heightening our acuity of vision, the most powerful is without doubt the use of

54. Woodger, J. H., *Proc. Aristot. Soc.*, *32* (1932), 110.

X-radiation. If within a solid there is pattern, we call the solid crystalline, and in fact there are very few solids which have not some degree of pattern within them. Now although the waves which give rise to human vision are about 10,000 times too coarse to show

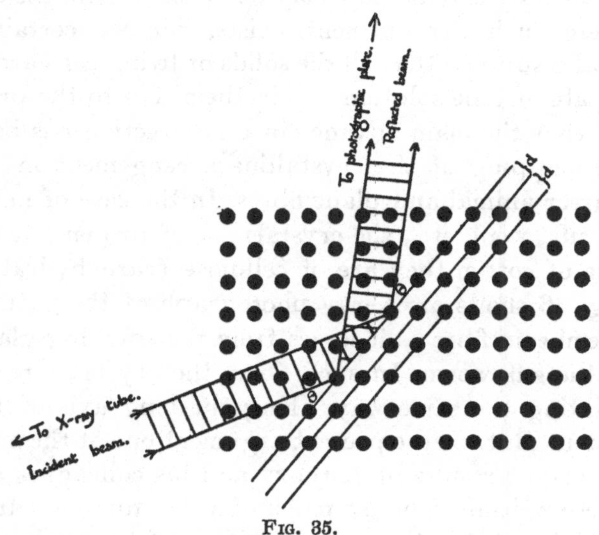

FIG. 35.

Diagram illustrating the reflection of an X-ray beam by a layer of molecules in the body of a crystal. (From Astbury.)

up any of these molecular patterns, it is possible to have recourse to X-rays, and by observing the patterns produced when a bundle of these is shot through a given solid, to draw deductions about the nature of the pattern within. The crystal, being a regular structure, reflects the X-rays in a regular way, and the result, when photographed, can be interpreted with the aid of the known laws of such reflection (Fig. 35).

It has been said that the crystalline state is the natural state of solid matter. The significance of this for biology is that there may be phases within the cell where such arrangement exists, for we certainly cannot suppose that all the solids of living cells are in a state of true solution within them. Up to the present time the main advance in this direction has been the mapping of the crystalline arrangement in the coarser animal and plant fibres. In the case of natural silk, wool, etc., the crystals are of protein, in the case of cotton they are of cellulose (carbohydrate). Fig. 36 shows an X-ray photograph of the natural silk fibroin fibre as it issues from the spinning-gland of the silk-worm. In such fibres the crystals are all pointing the same way, along the long axis of the system. But for the present argument one of the most important results of Astbury and his colleagues (in whose writings[55] lies so much of value for the future of biology) was their establishment of the chain-like nature of the protein molecules in such crystals. This

55. Astbury, W. T., *Trans. Faraday Soc., 29* (1933), 193; Astbury, W. T., and Street, A., "X-ray Studies of the Structure of Hair, Wool, and Related Fibres, I. General," *Phil. Trans. Roy. Soc., 230* (1931 A), 75; Astbury, W. T., and Woods, H. J., "X-ray Studies of Hair, Wool, and Related Fibres, II, the Molecular Structure and Elastic Properties of Hair Keratin." *Phil. Trans. Roy. Soc., 232* (1933 A), 333; *also* Astbury, W. T., *Fundamentals of Fibre Structure* (London, Oxford University Press, 1933); *see also* Jordan Lloyd, D., "Colloidal Structure and Its Biological Significance." *Biol. Rev., 7* (1932), 254.

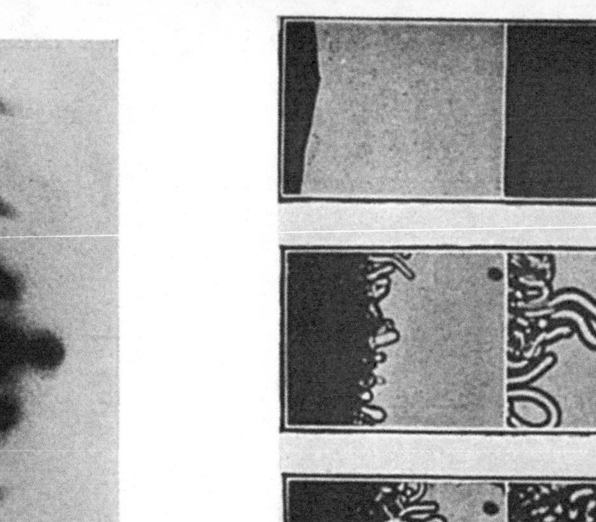

FIG. 36.

X-ray fibre-photograph of natural silk (fibroin.) (From Astbury.)

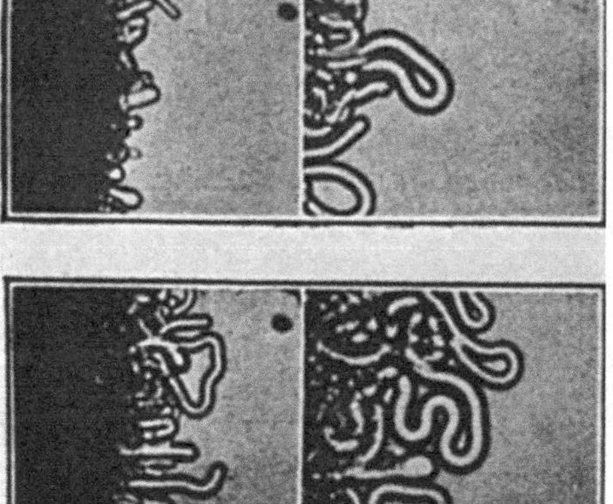

FIG. 44.

Six photographs from a cinematograph film showing the writhing movements ("myelin figures") produced by wetted lecithin. (From Leathes.)

fitted in at once with the previous knowledge of polypeptide structure, taking into account the setting of the valency bonds of the carbon and nitrogen atoms at or about the tetrahedral angle 109° 28′.

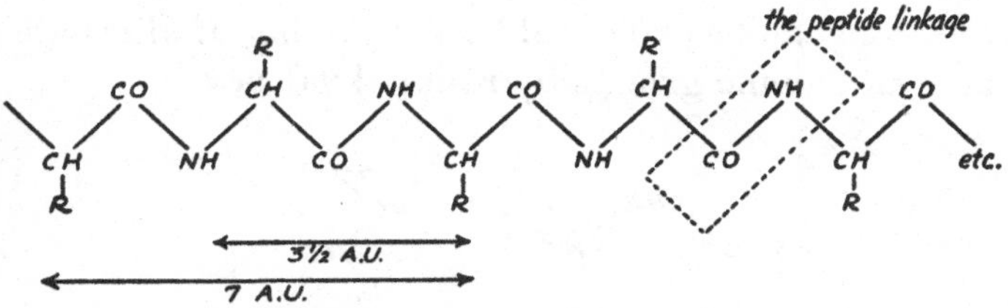

Thus each amino-acid of the polypeptide participates in the chain by means of its carboxyl and its α-amino carbon atoms, the rest going off as a side-chain, like a long and narrow garden from a road. The side-chains will not, of course, be of equal length, but if they follow each other in a definite periodic order, there will be a general regularity if the area under discussion is large enough. In the case of silk fibroin, the chain must almost entirely consist of alternating molecules of glycine and alanine.

The next advance originated from the circumstance that Astbury was a textile technologist, and therefore very much concerned about the shortening and lengthening of fibres. The molecular chains of silk fibroin and of cellulose were, he found, already extended to the maximum possible length, so that further extension could only take place by making the fibres slip over one another. In this there was no elasticity; the new length was permanent. Wool, however, shows remarkable elastic properties; in steam it can be pulled out to double its normal length, and yet return exactly to its original state if kept moist when the tension is removed. Astbury was able to show that this phenomenon is due to genuine *molecular contraction*, and can be explained by the joining of alternate keto and amino groups by residual valency

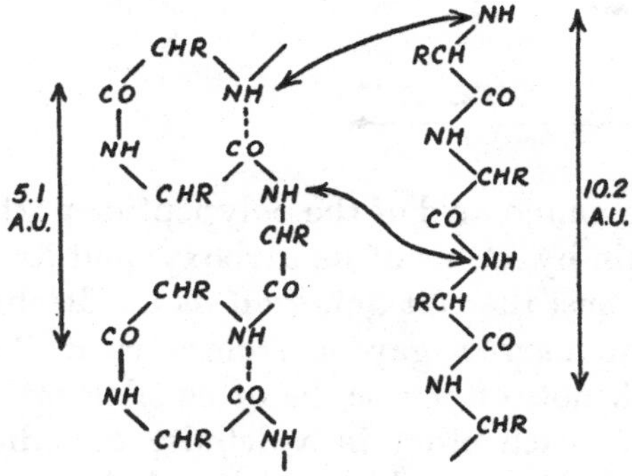

It will be noticed that the shortening is exactly 50%.

The importance of this work on the crystal structure of animal fibres can hardly be overestimated. Is not biology as a whole very largely the exploration of fibre properties? It is now practically certain in the case of muscle, for instance, that the contractile mechanism is essentially a molecular contraction. With this there ends a state of ignorance not much less intense than that which existed in the time of Descartes and Borelli. But we are especially interested in the egg-cell. It may be a far cry from the macroscopic fibres of Astbury to the intracellular webs of protein molecules of which Peters has spoken, but there is no reason to suppose that the one is less open to investigation than the other. And the subject has enormous importance in connection with the polarity and symmetry of the fertilised egg.

Embryologists have long been impressed with the capacity of eggs to develop normally after their contents have been thoroughly stratified by centrifugal force. The centrifuged amphibian egg, for instance, will show a mass of yolk-granules at the centrifugal pole, a cap of pigment at the opposite one, and an intermediate transparent protoplasmic layer. Conklin,[56] in particular, has drawn attention to the possible existence of a "spongioplasmic framework" not destroyed by centrifugation, in spite of the egg's "ballast" being movable through its meshes. There is much evidence, indeed, that the egg contains a "framework of viscid protoplasm which is so elastic or contractile that it recovers its normal form after distortion."[57] Such a web only conflicts with the fundamental results of Hardy at the beginning of the century if it is conceived of in too rigid a manner. The protein chains must be pictured rather as connected at many points by residual valencies and loose attachments, so that they can, as it were, snap back after disarrangement. Elsewhere Conklin[58] describes

56. Conklin, E. G., "Effects of Centrifugal Force on the Structure and Development of the Eggs of *Crepidula*," *Journ. Exp. Zool.*, *22* (1917), 311; *see especially* p. 351 ff.
57. Conklin, E. G., *loc. cit.*, p. 351.
58. Conklin, E. G., "Cellular Differentiation," contribution to Cowdry's *General Cytology* (1933), p. 562.

how the nuclei, centrospheres, yolk-inclusions, etc., of invertebrate eggs come back slowly to their usual places after the action of the pressure or centrifugal force has ceased. Wintrebert,[59] too, after much experience with the eggs of amphibia, speaks of the need for the hypothesis of a "cytosquelette" or "trame spongioplasmique" (Fig. 37, opp. p. 106).

One way of attacking the problem would be to ask whether the molecular shape of the protein molecules of the egg hyaloplasm is spherical (as respiratory pigment molecules, for instance, seem to be) or elongated. If the latter were true, it would be easier to imagine the interlacing web of protein chains. Longitudinal striations have been seen by Seifriz[60] in living amoeboid protoplasm by the aid of the Spierer oil-immersion dark-ground ultramicroscope. The most direct method, however, would be to prepare the protein of "yolkless" eggs (such as those of sea-urchins) and examine it, either with the ultracentrifuge of Svedberg[61] or with the rotating cylinder of v. Muralt and Edsall.[62] The two last-named workers studied the properties of muscle protein (myosin). Having sepa-

59. Wintrebert, P., "Les rotations chez l'oeuf de *Discoglossus*"—"trame résistante et elastique qui enclôt dans ses mailles et tient en place les divers matériaux ovulaires dans leur ordonnancement primitif." *Comptes Rend. Soc. Biol., 106* (1931), 439.

60. Seifriz, W., "The Structure of Protoplasm," *Sci. N. S., 73* (1931), 648.

61. Svedberg, T., "Sedimentation of Molecules in Centrifugal Fields," *Chem. Rev., 14* (1934), 1.

62. v. Muralt, A. L., and Edsall, J. T., "Studies in the Physical Chemistry of Muscle Globulin." *Journ. Biol. Chem., 89* (1930), 289, 315, 351.

rated it from the other constituents of muscle with the utmost care, they enclosed it in a space between two concentric revolving cylinders so that the shearing forces would orient the particles in a uniform direction. Then, observing it by means of optical apparatus through crossed Nicol prisms in a direction longitudinal to the cylinders, they found the "cross of isocline." This meant that the particles were all *rods* of exactly the same size, identical, in fact, with the crystals previously postulated to account for the double refraction of the muscle fibre. It would be most desirable to carry out similar work upon egg-proteins.

In this connection it is interesting that all eggs appear to contain two main classes of protein.[63] On the one hand there are the vitellins or ichthulins, phosphoproteins containing much serine phosphoric acid; on the other hand there are the livetins or thuicthins,[64] which are pseudo-globulins not unrelated to the myosin of muscle. Could it be that the former are primarily stores of nitrogen and phosphorus for the subsequent elaboration of the embryo's architecture, and that the latter (much less usually in amount) are the molecules constituting the "cytoskeleton"?

All the most reflective embryologists seem driven to the assumption of a "cytoskeleton," similar to that intuitively described by Peters (p. 135). Schleip, at the conclusion of his encyclopaedic work on embryonic determination, says: "In every attempt at the

63. Data in Needham, J., *Chem. Embryol.*, Pt. III, Sec. 1.

64. Needham, J., "A Note on Selachian Yolk-proteins," *Biochem. Journ.*, *23* (1929), 1222.

explanation of polarity and symmetry in the egg, some as yet unknown property of the protoplasm has to be introduced. To avoid giving it any new name, which could only be arbitrary and tentative, I will call it 'intimate structure'" (*Intimstruktur*).[65] Or again: "An intimate structure is present, i.e., a morphologically invisible, specific property of the cytoplasm, the manner of working of which we do not yet understand. To this must the difference between spiral and radial cleavage be referred, this determines too the direction of bilateral-symmetrical cleavage, and governs the planes of subsequent mitoses."[66] Przibram, in his work on the crystal-analogy, is still bolder, and does not hesitate to objectify the Thompsonian co-ordinates which were described in the first lecture. "Such systematic deformations," he says, "may most easily be explained on the assumption of an organic space-lattice. Just as by the substitution of radicals in eutopic series of crystals, deformations may be obtained because the effect is the same on all parts of the lattice, so the substitution of one protein combination for another would bring about a parallel change in the organic space-lattice."[67]

This is not nearly so extravagant as it sounds, if we remember the elastic properties of protoplasm. An entanglement or "brush heap" of interlacing crystal-

65. Schleip, W., *Die Determination der Primitiventwicklung* (Leipzig, Akad. Verlag., 1929), p. 850.

66. *Loc. cit.*, p. 856.

67. Przibram, H., *Die anorganische Grenzgebiete d. Biologie* (Berlin, Börntrager, 1926), p. 190; also *Aufbau mathematischer Biologie* (Berlin, Bornträger, 1923), p. 13.

line fibres or amino-acid chains is elastic; an emulsion of spherical particles is not.[68]

The "cell-skeleton" is, however, by no means easy to discover. Among the interesting recent attempts to demonstrate it are those of Howard and Vlès. Howard studied the apparent viscosity of sea-urchin egg protoplasm at various rates of shear by observing the rate of movement of granules under centrifugal force.[69] No unequivocal plasticity was shown by the egg at rates of shear so low that the granules were moving at velocities comparable to their own migration velocities (i.e., their velocities of return to place after centrifuging). No continuous structure, it was concluded, was present which could significantly affect diffusion, nor, *a fortiori*, be the basis of polarity. But it is uncertain whether what we are looking for could be demonstrated by any centrifuge method. Vlès, again, has studied the moduli of elasticity and

68. *See* the admirable discussion of protoplasmic structure by Seifriz, W., "The Physical Properties of Protoplasm," contribution to *Colloid Chemistry, Theoretical and Applied,* Vol. II (Biol. and Med.), ed., J. Alexander (New York, 1928), p. 403, *especially* p. 441 ff., also further discussion in *Bot. Rev., 1* (1935), 18. The picture of cellular organisation given by him is closely similar to that of Peters.

"We can regard protoplasm," says Seifriz (*Amer. Nat., 60* [1926], 124), "as essentially a protein-like substance, fibrous in structure, which is permeated by an emulsion of food material. The physical properties which so fundamentally characterise protoplasm, such as elasticity, rigidity, and imbibition, exist only in virtue of its fibrous structure."

69. Howard, E., "The structure of protoplasm as indicated by a study of the apparent viscosity of sea-urchin eggs at various shearing forces." *Journ. Cell. and Comp. Physiol., 1* (1932), 355. The paper contains references to earlier thought of embryologists about the spongioplasm.

rigidity in the sea-urchin egg as it becomes a sphere again after deformation to a sausage-like shape.[70] His results do not exclude the possibility of the structure we are envisaging. On the other hand, the extraordinary tendency of protozoa to swim in spirals, some right and some left, leads Schaeffer[71] to postulate a stereochemical difference in the protein framework, and Rand and Hsu[72] have a passage of particular interest. Describing the motion of amoeboid cells, they say: "In watching the nucleus one is perplexed by the fact that it is so freely movable within limits which are in no way visibly defined. Apparently freely immersed in a labile and actively flowing protoplasm, why is its position not completely at the mercy of the currents? It must possess some highly elastic anchorage. It is as if a slightly buoyant sphere, immersed in a strongly flowing stream, were anchored by relatively slender and extremely elastic cables."

I will illustrate my aphorism that biology is largely the study of fibres by further examples. In earlier times there was much discussion about what

70. Vlès, F., "Recherches sur une déformation mécanique des oeufs d'oursin." *Archives de Zool. exp. et gen., 75* (1933), 421.

71. Schaeffer, A. A., "On Molecular Organisation in Amoeban Protoplasm," *Sci. N. S., 74* (1931), 47. cf. the possibility that the origin of the sinistrality and dextrality exhibited by certain molluscan eggs in their cleavage (Boycott, A. E., Garstang, S., and Diver, C., *Journ. Genetics, 15* [1925], 113) may be referable to stereochemical properties of the protein molecules composing them.

72. Rand, H. W., and Hsu, S., "Concerning Protoplasmic Currents Accompanying Locomotion in *Amoeba*," *Sci. N. S., 65* (1927), 261.

was called the inadequacy of the cell-theory of development.[73] But whereas the discussion then mainly turned on the meaning of the histological evidence, i.e., the existence and nature of cell-boundaries, in our day the importance of the cell has fallen a little into the background because of the increasing tendency of embryologists to think in terms of *regions*. The process of determination, for instance, under the influence of an organiser, takes little account of single cells and operates, or is thought of as operating, in a more statistical way. The discussion in Russell's recent book[74] thus found little echo among biologists. But recently Moore[75] has greatly revivified the question by demonstrating with the micro-manipulator the existence of cell-bridges in the early development of invertebrate eggs, and by suggesting theories of gastrulation which necessitate the help of these cell-bridges. Here is a fundamental field for the application of X-ray analysis. What would we not give to know the crystal structure of the protein

73. Perhaps the two classical papers are: Sedgwick, Adam, "On the Inadequacy of the Cellular Theory of Development," *Quart. Journ. Mic. Sci., 37* (1895), 87, and Whitman, C. C., "The Inadequacy of the Cell-theory of Development," *Journ. Morphol., 8* (1893).

74. Russell, E. S., *The Interpretation of Development and Heredity* (Oxford, Clarendon Press).

75. Moore, A. R., "Fertilisation and Development without Membrane Formation in the Egg of the Sea-urchin," *Protoplasma, 9* (1930), 9, 18. "On the Invagination of the Gastrula," *Protoplasma, 9* (1930), 25. Moore, M. M., "On the Coherence of the Blastomeres of Sea-urchin Eggs," *Archiv. f. Entwicklungsmech., 125* (1932), 487. Whong, S. H., "On the Formation of Cell-bridges in the Development of the Sea-urchin Egg," *Protoplasma, 12* (1931), 123.

chains of these cell-bridges? Is it not significant, in view of what we now know of molecular contractility, that they possess considerable elasticity? Contractility, indeed, may play a far larger rôle in development than most embryologists suppose. Let us only remember the blastodermal ripplings of teleostean embryos (Amemiya) and the extremely mysterious cell-streams which are basic for amphibian and avian development. Then in insect eggs there is the curious phenomenon of blastokinesis, and even more important, the wave of yolk-contraction at the differentiation-centre (Seidel).

Lastly there are the fibres of the connective tissue ground-substance in later development. Weiss[76] has shown that in tissue-culture cells will tend to grow along in the direction of stresses orienting the micelles of the colloidal medium. Fig. 38 (opp. p. 106) taken from Snessarev shows a micro-photograph of the outer layer of the exocœlom of a chick embryo of 72 hours' incubation. The primitive connective-tissue fibrils can plainly be seen. Applying Weiss's result to embryonic development *in vivo* it is evident that the orientation of ultra-structure, protein rods, etc., in the ground-substance (e.g., the mesostroma filling the spaces be-

76. Weiss, P., "Functional Adaptation and the Rôle of Ground-Substances in Development." *Amer. Nat., 67* (1933), 322; *see also* Needham, J., *Chemical Embryology*, pp. 566, 1012. The papers of G. A. Baitsell contain much of value:—*Journ. Exp. Med., 23* (1916), 739; *Amer. Journ. Physiol., 44* (1917), 109; *Amer. Journ. Anat., 28* (1921), 447; *Quart. Journ. Mic. Sci., 69* (1925), 571; *Amer. Rev. Tuberc., 21* (1930), 593, *29* (1934), 587.

tween the germ-layers), quite apart from any visible fibres, may be of enormous importance in the directive processes of differentiation.

Many cases, of course, are known to embryologists where a non-cellular tract is first laid down, and afterwards the cells migrate into it (cf. the uveal tract—Shearer). As Bauer says, "Erst später wachsen in den von kollagen Fasern erfüllten Raum die Mesenchymzellen hinein."[77] And, "Die Intercellularsubstanz wird von den Zellen gebildet, steht unter deren dauernden Einwirkung und bildet mit ihnen zusammen ein lebendiges Ganzes."

We do not of course yet know the normal equilibrium configuration of most of these fibres. We could guess, however, that the Weissian fibres are usually contracted, like the chains of wool, while the cell-bridges of echinoderms and the yolk-fibres of insects are usually expanded, like the chains of silk.

I cannot leave the subject of fibres and microfibrils without referring to a type of living matter at first sight somewhat far removed from the animal embryo—the plasmodium or slime-mould. Such a mass of protoplasm will flow through cotton-wool according to an observation of Arthur Lister in 1888: "I placed some wet cotton-wool in front of the still dingy plasmodium; this it readily penetrated, and afterwards emerged possessing its normal yellow colour, leaving the wool charged with spores and

77. Bauer, C., *Zeitschr. f. mik. Anat. Forschung, 35* (1934), 362.

other debris."[78] Recently Moore[79] has pushed further the analysis of protoplasmic structure along these lines. *Physarum polycephalum,* a slime-mould, was not killed by being ground in a mortar with quartz sand, but was killed if finely crushed glass was used instead. Both soft and hard filter papers permitted the passage of the plasmodium, and it readily passed through parchment, where the pores were of an average diameter of 5×10^{-5} mm. Yet if pressure was applied, as by forcing it through textiles or filter paper, the protoplasm was destroyed at far larger pore-sizes. Thus to be squeezed through fine silk (pore-size 5×10^{-2} mm.) was lethal for it, and pores as large as 0.25 mm. had to be used before it could be safely pressed through. "These results suggest," says Moore, "that the plasmodium contains long threads of living material essential to its existence, and if these be broken too short, life is impossible." From the rough data just given such micro-fibrils would be 5×10^{-5} mm. in diameter and 2000 times as long, i.e., about 100 times the diameter of the protein chains envisaged by Peters. However, the general suggestiveness of these experiments cannot be questioned.

The aspect of molecular pattern which seems to have been most grossly underestimated in the consideration of biological phenomena is that found in

78. Lister, A., *Ann. Bot.,* *2* (1888), 2.

79. Moore, A. R., *Sci. Rep. Tohoku Imp. Univ. (Biol.),* *8* (1933), 189; and *Proc. Soc. Exp. Biol. and Med.,* *32* (1934), 174. *See also* Seifriz, W., *Rev. Gén. de Bot.,* *46* (1934), 200, and *Physics,* *6* (1935), 159. The acid of injury must be taken into consideration in evaluating Moore's results on pore-size.

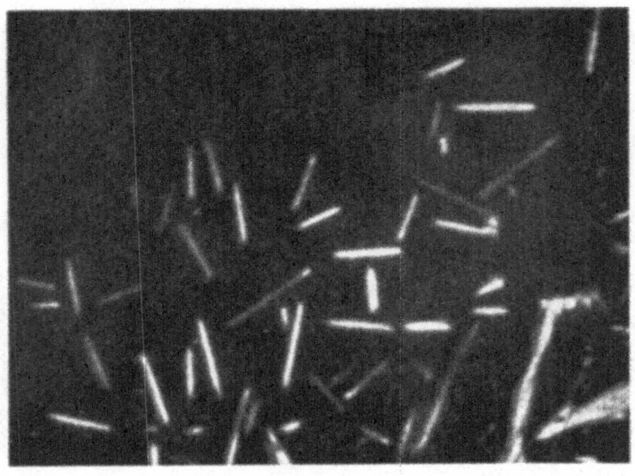

Fig. 39.

Paracrystalline para-azoxy-benzoic-acid-ethylester under crossed Nicol prisms. (Vorländer, after Rinne.)

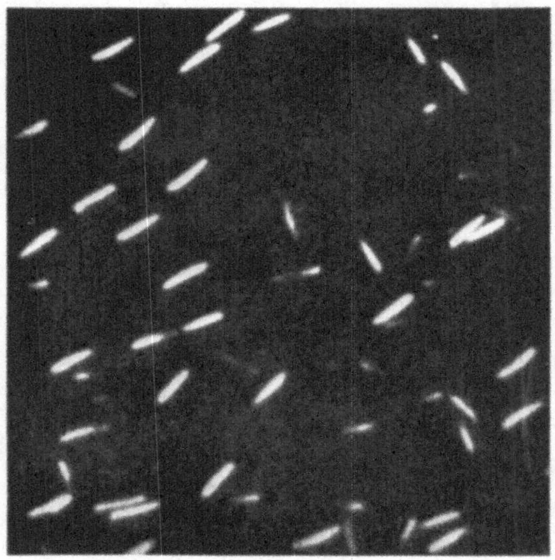

Fig. 40.

Spermatozoa of the squid, *Sepia,* under crossed Nicol prisms. (Schmidt, after Rinne.)

liquid crystals.[80] It is likely that progress in the discovery of the "leptonic fibres"[81] of which we have been speaking will be delayed until more knowledge is available of the peculiar form of order which liquid crystals exhibit. Liquid crystals, it is to be noted, are not important for biology and embryology because they manifest certain properties which can be regarded as analogous to those which living systems manifest (models), but because living systems actually *are* liquid crystals, or, it would be more correct to say, the paracrystalline state undoubtedly exists in living cells. The doubly refracting portions of the striated muscle fibre are, of course, the classical instance of this arrangement, but there are many other more striking instances, such as the cephalopod spermatozoa studied by Schmidt[82] (Figs. 39, 40). Even

80. Cf. Rinne, F., "Investigations and Considerations Concerning Paracrystallinity," contribution to Faraday Society Symposium on *Liquid Crystals and Anisotropic Melts* (London, 1933); *also Grenzfragen des Lebens* (Leipzig, Quelle & Meyer, 1931).

81. This expression is taken from the suggestion of Rinne (Faraday Society Symposium, *Liquid Crystals*, p. 1029) that for international purposes parallel words are needed for the German "Feinbaulehre," "Feinbau." He suggests "Leptology," "leptonic," from λεπτός, delicate, fine. In accepting these words, I follow the use of W. Lubosch, in what is surely the most philosophical treatise on anatomy ever written—*Outlines of Scientific Anatomy* (London, 1928). The work of E. S. Bauer is along this direction; see his *ТЕОРЕТИЧЕСКАЯ БИОЛОГИЯ* (Leningrad, 1935), *also* contributions to *Probleme der Theoretischen Biologie* (Leningrad, (1935); and *Proc. XVth. Internat. Congress of Physiology*, Leningrad, 1935, p. 24. Bauer suggests that a part of basal metabolism is work done in the deformation of the leptonic fibres.

82. Schmidt, W. J., *Zool. Jahrb.*, *45* (1928), 177, *also* Cohen, *Die Bausteine des Tierkörpers in polarisiertem Lichte* (Bonn, 1924).

in the egg there are the anisotropic yolk-platelets.[83] And it is very probable that the paracrystalline state exists in many phases in the cell whose scarcity or position has so far rendered them immune from investigation and invisible when the living cell is placed between crossed Nicol prisms. As Rinne points out, the Langmuirian oriented film of fatty acid on water may be regarded as paracrystalline, and this is certainly a close relation of the oriented films within the cell or the egg.

When a solid is heated, our ordinary conceptions are that it melts when a certain temperature is reached, passing immediately into the state of an isotropic liquid, where the molecules are arranged at random according to kinetic theory. This, however, is much too simple for many substances, which on the contrary pass through a succession of "mesoforms," conditions where a certain amount of liquid-like flow is possible, and yet where there is within the liquid a definite and regular arrangement of molecules. These mesoforms are now classified as follows. Remembering that all paracrystalline substances have very elongated molecules several different cases may arise (Fig. 41). In I the molecules are completely at random; this is the true isotropic liquid state. In II all the molecular axes are parallel, but the centres of the molecules are as irregularly arranged and as free to move as in the first case. This is called the *nematic*

83. Schmidt, *loc. cit.*, pp. 455 ff. They are the granules of lecitho-vitellin (ichthulin or ranovin)—*see* Needham, J., *Chemical Embryology*, Pt. III, Sect. 1.

I. Isotropic liquid. Neither orientation nor periodicity.

II. Nematic. Orientation without periodicity. Above, viewed perpendicular to molecular axis; below, parallel to molecular axis.

III. Normal Smectic. Orientation and molecules in equispaced planes with no internal periodicity. Above, viewed perpendicular to molecular axis; below, parallel to molecular axis.

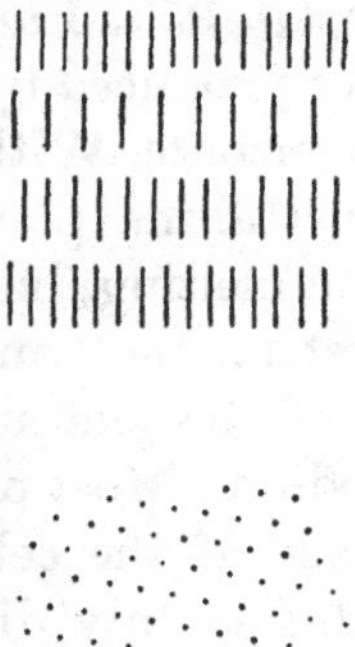

IV. Low Temperature Smectic. Each layer with two dimensional periodicity but unrelated to other layers.

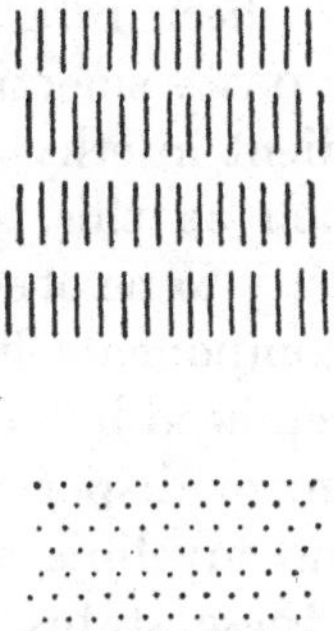

V. Crystalline.
(*a*) Smectogenic. (*b*) Nematogenic.
Complete orientation and three dimensional periodicity.

Fig. 41.

Chart to show the different kinds of paracrystalline states. (From Bernal.)

state. In III also the molecular axes are parallel, but the centres of the molecules have lost one degree of freedom and are now restricted to a set of regularly spaced parallel surfaces. This is the *smectic* state. In IV the smectic state is still further ordered by having the molecules regularly arranged within each layer. V gives the conditions in the crystalline solid state. The molecules are parallel and their centres form a regular three-dimensional network. A smectogenic crystal has its molecules arranged in planes (Va); a nematogenic crystal has them arranged so as to interleave each other (Vb). In addition to these states there may be others, and it is interesting that one has long been recognised, the most complicated of all, called *cholesteric*, since it is typically shown by the fatty acid esters of cholesterol. The significance of this fact has already been referred to (p. 95). Fig. 42 shows the oily or pasty smectic states produced by a temperature gradient in ethylazoxybenzoate. With pure substances such as this, paracrystalline phenomena are most easily to be observed in melting, but with two or more components in a system, the temperature factor is replaced by changes in the concentration of the solvent or dispersing medium. Most of the protein, fat, and myelinic substance of the cell probably exists in these states, but this is only directly visible when all the molecules are oriented in enormous swarms in one direction, as in muscle fibrils. The paracrystalline state seems the most suited to biological functions, as it combines the fluidity and

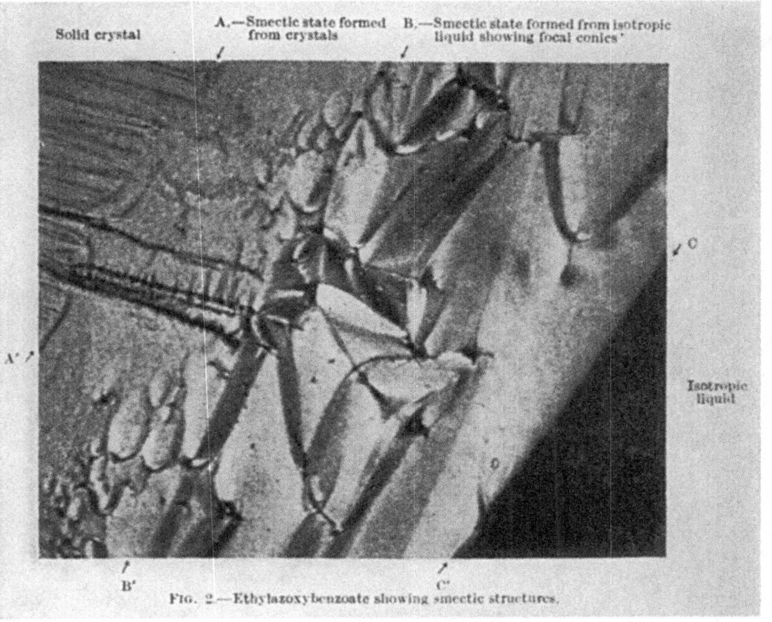

FIG. 2.—Ethylazoxybenzoate showing smectic structures.

FIG. 42.

Ethylazoxybenzoate showing smectic structures. (From Bernal.)

diffusibility of liquid while preserving the possibilities of internal structure characteristic of crystalline solids.

This has been very well put by Bernal. "The biologically important liquid crystals," he said, "are plainly two or more component systems. At least one must be a substance tending to paracrystallinity and another will in general be water. This variable permeability of liquid crystals enables them to be as effective for chemical reactions as true liquids or gels as against the relative impenetrability of solid crystals. On the other hand, liquid crystals possess *internal structure* lacking in liquids, and *directional properties* not found in gels. These two properties have far-reaching consequences. In the first place, a liquid crystal in a cell through its own structure becomes a *proto-organ* for mechanical or electrical activity, and when associated in specialised cells (with others) in higher animals gives rise to true organs, such as muscle and nerve. Secondly, and perhaps more fundamentally, the oriented molecules in liquid crystals furnish an ideal medium for catalytic action, particularly of the complex type needed to account for growth and reproduction. Lastly, a liquid crystal has the possibility of its own structure, singular lines, rods and cones, etc. Such structures belong to the liquid crystal as a unit and not to its molecules, which may be replaced by others without destroying them, and they persist in spite of the complete fluidity of the substance. They are just the properties to be required for a degree of organisation between that of the continuous substance, liquid or crystalline solid, and even the simplest living cell."[84]

It has been known for a long time that many constituents of the living cell manifest remarkable interfacial properties when in contact with water (lipoids and sterols); this knowledge long antedates our information on the crystalline character of protein chains. The "myelin figures" produced by wetted lecithin are classical. They are undoubtedly connected with the solubility of the phosphoric acid and glycerol groups in the molecule and the insolubility of the hydrocarbon chains of the fatty acids. First spheres with optically positive radii are formed, and from these there grow out excrescences and long sluggishly writhing masses, often with remarkable turns, twists, spiral wrappings, and swellings. As the water content is increased the double refraction falls to a low value (Fig. 43). Leathes,[85] who pleads for the view that protein structure alone will not be adequate for the understanding of intracellular structure, has filmed these myelin forms with the cinematograph un-

84. Bernal, J. D., Remarks on F. Rinne's paper, Faraday Society Symposium, *loc. cit.*, p. 1082; italics mine. cf. also the paper of J. W. McBain on structure in amorphous and colloidal matter; *Journ. Chem. Educ.*, *6* (1929), 2115.

85. Leathes, J. B., Croonian Lectures, "The rôle of fats in vital phenomena," *Lancet* (1925), 803, 853, 957, 1019. The general conclusions of Leaths are in close agreement with the point of view here put forward. Protoplasm, he writes, is not miscible with water until it disintegrates, because, though liquid, it has structure. The solid phases are like "expanded seaweed floating in the sea."

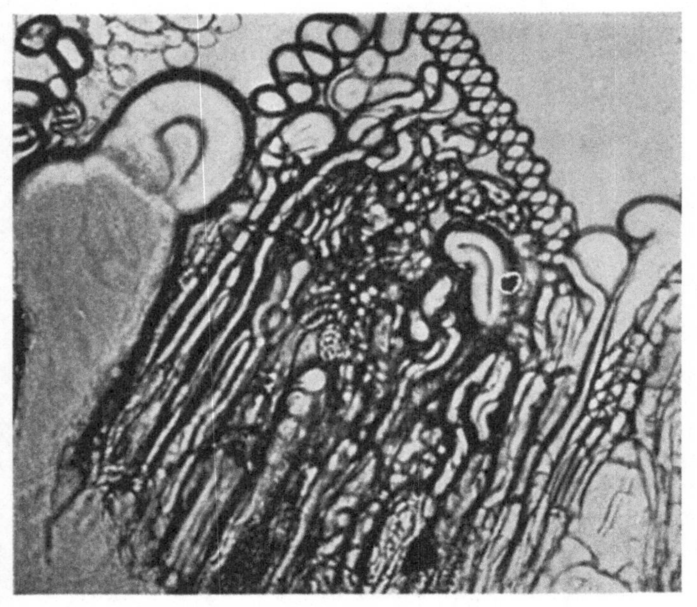

Fig. 43.

Morphological forms produced by wetted lecithin. (From Leathes.)

der many different conditions. An example of his pictures is given here (Fig. 44 opp. p. 144).[86]

It has been pointed out[87] that the mesoforms can be defined in terms of dimensions. A solid is rigid in three dimensions, a smectic mesoform in two, a nematic mesoform in one, an isotropic liquid in none. This can easily be appreciated by reference to Fig. 41. On heating, melting occurs in the three dimensions in turn; the three different melting-points being the temperatures at which the three vectorial forces of adhesion are overcome by thermal agitation. This formulation, though not quite accurate, draws attention to the fact that the basic observation on mesoforms is that several melting-points succeed one another in the transition from solid to liquid. Now there is a similarity here between these successive stages of dimensional rigidity, and the very curious phenomena seen in the determination of limb-buds in amphibia.[88] When discs of the outer wall of the body, representing the buds of the future limbs, are transplanted at a certain stage, it is found that the original anterior edge of the bud always produces the preaxial part of the limb. So a limb-bud of the left side, planted the right way up on the right side of the embryo, will de-

86. The optical and other properties of the myelin figures as studied in vitro are closely similar to those of the sheaths of intact nerves, *see* Rinne, F., *Kolloid Zeitschr.*, *60* (1932), 288.

87. e.g., by Lawrence, A. S. C., and Rawlins, F I. G., *Sci. Progr.*, *28* (1933), 339.

88. The following rapid condensation of the experiments of Harrison on limb polarity is taken from de Beer, G. R., *Experimental Embryology* (Oxford, Clarendon Press, 1926), p. 74.

velop into a limb with the elbow pointing forward instead of backward, if it is a forelimb. On the other hand, if the disc is rotated so that the original dorsal edge of the bud is ventral, the original anterior edge will be anterior again, and a normal limb develops. Therefore the dorso-ventral axis can be inverted without producing rearrangement, but not so the antero-posterior axis. The medio-lateral axis can also be inverted at this stage with impunity, i.e., it does not matter whether the limb-bud is attached to its new situation proximally or distally. This means that the dorso-ventral and medio-lateral axes are still plastic, while the antero-posterior axis is determined. Later on they also become determined. Here, then, we have successive stages of dimensional determination. Just as the liquid crystal passes through stages of rigidity in one, two, and three dimensions, so the limb-bud passes through stages of determination in one, two, and three dimensions. The analogy here may, of course, be purely superficial, but it is sufficiently striking to warrant attention. To quote Hardy again, when one has long contemplated a problem toward the solution of which nothing could be advanced, even the slightest clue is welcome.

To recapitulate, a careful consideration shows us that the fields of morphology and biochemistry are not so sundered as is often supposed. Organising relations are found at the molecular level and at the colloidal and paracrystalline level as well as at the anatomical level. Hardy's work, far from showing that no structure existed in the cell, showed on the contrary how subtle it must be. Although we are still in the earliest historical stages of any far-reaching organisation-calculus, we can yet see that biological order, like (but very much more complicated than) crystal order, is a natural consequence of the properties of matter, and one characteristic mode of their manifestation.

This has been well put by Sapper. "We now stand," he says, "before a problem which the supporters of the Gestalt-theory have hardly yet answered, namely, how is the origin of pattern (*Gestalt-charakter*) in material objects in general and living things in particular, to be explained? Is it not indeed inconceivable that properties should be found in a material complex, which are *not* the result of the summation of the properties of the components? Are we really forced to the assumption of some supra-material, hyper-individual, factors, in order to account for the appearance of the qualitatively new in the organised patterns? In my view there is only one way to picture the organisation of a material complex without having recourse to such assumptions; and that is to assume that the qualitatively new in the pattern derives from the properties of the elements involved, but that certain of these properties can only come into operation in connection with certain specific stages of complexity. There is of course no proof available for demonstrating the rightness of this viewpoint. But it will not be denied that it describes the facts in the simplest way and has the advantage of agreeing with the analogy from the social life of man [e.g., combination of musicians in an orchestra, etc.]. If one disagrees with it, one has the choice, *either* of seeking to contest the facts of the existence of non-additive properties in complex patterns, *or* of regarding them as fundamentally inexplicable and unintelligible."[89] Or, as Meyerhof[90] says "One must take the view that the formative and functional forces, manifested by matter in living beings, are already *latent* there."[91]

In H. M. Tomlinson's novel *All our Yesterdays*, the mycological vicar soliloquises on wars and civil tumults: "The trouble in our system comes out badly now and then like some rare toadstools do at times. I was looking up a fungus just now before I came out.

89. Sapper, K., *Philosophie d. Organisches* (Breslau, Hirt, 1928), pp. 85, 88.

90. Meyerhof, O., *Naturwiss.*, *22* (1934), 312.

91. It is obvious that we are here verging on a subject with a large literature of its own, the theory of Emergent Evolution. I will only refer those interested to a paper by Morgan, C. Lloyd, "A Concept of the Organism, Emergent and Resultant," *Proc. Aristot. Soc.*, *27* (1927), 141. Emergent Evolutionism seems to hold that the specific properties of wholes are conferred on them over and above the specific properties of their parts by a continuous Creativity acting from outside. Dialectical Materialism, as suggested in the first lecture, seems to hold that the specific properties of wholes result from properties of the parts which are invisible or latent in isolation. The former view, which is not dissimilar to the deist conception of the "general concourse" whereby God continuously upholds his creation, seems suited to a religious world-outlook. The latter view seems equally suited to a scientific world-outlook. The problem of which view is correct, if it is a problem, appears to be extra-scientific and quite insoluble.

Nobody remembers having seen it before, but there it is, all of a sudden—thousands of it. Presently, it will disappear for years and years and we'll forget it. The blessed spores wait in the soil, though you don't know it. They make a display whenever the temperature and moisture are precisely right together, I suppose, and that doesn't often happen." In some such way, probably, it is best to conceive of the origin of life on earth—when cosmic conditions permit, matter produces in actuality what it has always had within it in potentiality.

In conclusion, I would refer to the perplexities of Driesch, at the beginning of the century, discussed in the previous lecture. A great deal of water has flowed under the bridges since then. We no longer feel the necessity which he felt to place the "intensive manifoldness" of the egg outside space-time. In 1895 very little was known of the complexities of the colloidal state, next to nothing about molecular orientation at interfaces, and nothing at all about the biological significance of paracrystals. Even the non-symbolic status of organic structural formulae was in doubt. The possibilities inherent in the field-concept were quite unexplored, and the existence of physiological gradients had not been discovered. Practically nothing was known about the relational properties of development as manifested through the action of organisers or evocators of various grades. The potentialities of the protein chain, and the phenomena of molecular deformability and contractility were unguessed at, and there was no hint of the exploration of solid bodies by X-ray analysis. These many and great advances give us every promise of a profounder insight into the nature of organic form. To abandon the quest at this stage would surely be the height of folly.

INDEXES

LIST OF LECTURES QUOTED

Croonian	London	Spemann	82 ff.
do.	London	Fletcher & Hopkins	140
do.	London	Leathes	162
do.	London	Keilin	124
Donnellan	Dublin	J. S. Haldane	9 ff.
Gifford	St. Andrews	Driesch	51 ff.
Goulstonian	London	Dodds	88
Guthrie	London	Hardy	137
Harben	London	Peters	137 ff.
Herter	Baltimore	Hopkins	140
Purser	Dublin	Hopkins	31

ADDENDUM

Fig. 45.

View of a marshalling yard (London & North Eastern Railroad) illustrating by analogy the concept of restriction of potentiality by multiple bifurcation. The photograph is taken from above the "hump." Up to this the freight wagons are pushed, and from it they run down individually over electric retarders to a number of alternative sidings, where they are ready to be despatched to a fresh destination. The top of the "hump" corresponds to the totipotent or maximally unstable condition of the egg.

INDEX

Printed in the United States
By Bookmasters